The Language of Time

International Library of Philosophy and Scientific Method

EDITOR: TED HONDERICH
ADVISORY EDITOR: BERNARD WILLIAMS

A Catalogue of books already published in the *International Library of Philosophy and Scientific Method* will be found at the end of this volume.

The Language of Time

by

RICHARD M. GALE
Associate Professor of Philosophy
University of Pittsburgh

LONDON
ROUTLEDGE & KEGAN PAUL
NEW YORK : HUMANITIES PRESS

First published 1968
by Routledge *&* Kegan Paul Limited
Broadway House, 68–74 Carter Lane
London, E.C.4

Printed in Great Britain
by Richard Clay *(The Chaucer Press)* Ltd
Bungay, Suffolk

SBN 7100 3637 x

CONTENTS

With all my love

to

MAYA

PREFACE

WHATEVER merits this book might have are due in large part to the stimulation and help I have received from other people. First I must mention the inspiration and encouragement I received from four former teachers: Professor Loyd Easton of Ohio Wesleyan University and Professors William Barrett, Sidney Hook and Milton Munitz of New York University. In developing my ideas about time I have learned a lot from discussion and/or correspondence with: Raziel Abelson, Bruce Aune, Kurt Baier, Nuel Belnap, Richard Burian, Charles Chihara, Arthur Danto, Fred Dretske, Gertrude Ezorsky, Edmund Gettier, Gilbert Harman, John King-Farlow, Morris Lazerowitz, Bernard Mayo, Milton Munitz, Stephen Norris, Nicholas Rescher, John Robison, Jay Rosenberg, Thomas Schwartz, Daniel Sedey, Wilfrid Sellars, J. J. C. Smart, Irving Thalberg, Frank Tillman and Martin Tweedale. I owe a particular debt to my friend and colleague, Professor Adolf Grünbaum. By word and deed he has shown me what it is to philosophize seriously about time. In addition to learning a lot from his monumental work, *Philosophical Problems of Space and Time*, I have had the privilege of auditing two of his courses. Numerous improvements were made in an earlier draft of this book due to the helpful suggestions of Ted Honderich, the Editor of the International Library of Philosophy and Scientific Method.

This work was begun during 1960–61 under a Samuel S. Fels Foundation Fellowship, was further pursued during 1964–65 under National Science Foundation grant number 387, and was brought to completion in 1966–67 under a Post-Doctoral Andrew Mellon Fellowship at the University of Pittsburgh. I hope that this book in some small way bears out the confidence that these foundations have shown in me.

Some of the chapters of this book draw upon previously published articles of mine, among which are: 'Endorsing Predictions', *Philosophical Review*, LXX (1961); 'Can a Prediction "Become True"?', *Philosophical Studies*, XIII (1962); 'Tensed Statements', *Philosophical Quarterly*, 12 (1962); 'A Reply to Smart, Mayo, and Thalberg on "Tensed Statements",' *Philosophical Quarterly*, 13 (1963); 'Is It Now Now?', *Mind*, LXXIII (1964); 'The Egocentric Particular and Token-Reflexive Analyses of Tense', *Philosophical Review*, LXIII (1964); 'The Generality of Predictions' (co-authored with Irving Thalberg), *Journal of Philosophy*, LXII (1965); 'Falsifying Retrodictions', *Analysis*, 25 (1965); 'Existence, Tense and Presupposition', *Monist*, 50 (1966); 'Why a Cause Cannot Be Later Than Its Effect', *Review of Metaphysics*, XX (1966); 'McTaggart's Analysis of Time', *American Philosophical Quarterly*, 3 (1966); 'Pure and Impure Descriptions', *Australasian Journal of Philosophy*, 45 (1967); *The Philosophy of Time*, edited by R. M. Gale (New York, Anchor Doubleday Books: 1967). I am indebted to the publishers of these journals and books for permission to reprint some of the material contained in the above publications. In most cases these former publications of mine have been significantly altered in both style and content so as to eliminate certain confusions and errors. Having no doubts about induction, I am reasonably certain that this book will also contain numerous errors. The best that I can hope for is that my errors will be helpful to those who are labouring on the problem of time.

I am indebted to Aaron Rosenthal for preparing the index and to Nancy Hawkins for proof reading the manuscript.

Part One

McTAGGART'S PARADOX

I

McTAGGART'S PARADOX AND THE LANGUAGE OF TIME

THE aim of this book is to clarify certain large-scale features of our ordinary or common-sense concept of time. Such clarification should be welcome not only for its own intrinsic interest but also for the light it can shed on philosophical problems and paradoxes about time, of which there is no shortage. This concern with the *common-sense* concept of time will seem misguided to some scientifically oriented philosophers who look upon common sense as a repository of confusions and outmoded empirical beliefs. For what has been more discredited by modern science, they would charge, than the common-sense concept of time. But before we can assess the truth of this charge we must be clear as to the nature of this concept: thus the need for clarifying this concept.

The method which will be employed in achieving this clarification is that of linguistic analysis or ordinary language philosophy. What this calls for is a study of the language of time in its everyday employment. No attempt will be made to justify this method of doing philosophy; for the only adequate justification for a method of doing philosophy is to put it to work and see how much clarification and illumination it helps us to achieve. It is hoped that when the task of articulating the use of ordinary temporal discourse is completed our common-sense concept of time will have been clarified, and, even more hopefully, that then there will remain nothing which needs to be further clarified by the adoption of some other method of doing philosophy, such as phenomenology or rational reconstruction.

The method of linguistic analysis does not seek to supply us with a definition of 'time' or produce some verbal formula in which the meaning of every temporal expression will somehow be telescoped. Anyone who conceives of the problem of time as one of defining 'time' will invariably wind up lamenting with St. Augustine: 'What, then, is time? If no one asks me, I know; but if I wish to explain it to one who asketh, I know not.' It will be instructive to diagnose the source of this perplexity, since it will indicate a more fruitful way of clarifying our concept of time.

Augustine's perplexity is due to the fact that we both have an experiential awareness of time and know the correct use of temporal language but are mysteriously reduced to silence when we try to verbalize this understanding. Any definition that we might propose will suffer from circularity, since it will have to employ some temporal expression in the *definiens*. That 'time' is verbally or analytically indefinable cannot be conclusively demonstrated, but it can be supported inductively by showing that all definitions which have been proposed so far are circular. As a start, we might look at three famous definitions of 'time' proposed by past philosophers. Aristotle's definition of 'time' as the 'number of movement in respect of the before and after' employs in the *definiens* the temporal expressions 'before' and 'after', as well as 'movement', which is defined in terms of time. Plotinus' definition of 'time' as 'the life of the Soul in movement as it passes from one stage of act or experience to another' is circular for the same reason, since 'movement', as well as 'passes', occurs in its *definiens*. Augustine's claim that there is only: 'a present of things past, memory; present of things present, sight; present of things future, expectations', is circular if intended as a definition of 'time', because 'memory', 'sight' and 'expectation' cannot be defined without reference to a *past*, *present* or *future* event which is respectively remembered, seen or expected.

Augustine's perplexity is not due solely to the *verbal* indefinability of 'time', for, if it were, he just as well could have asked in anguish, 'What, then, is yellow? . . .' The difference between 'time' and verbally indefinable words such as 'yellow' is that yellow, unlike time, can be pointed to, and therefore admits of a straightforward ostensive or demonstrative definition. Obviously, there is nothing we can point at and say, 'This is time.' We even cannot ostensively define the past, present and future, of which time is composed according to Augustine. For, by definition, the past and future cannot now be pointed at. The present, which seems to be more available, turns out to be equally elusive; for there is no sensible property or thing which we can point to and say: 'This is the present.' Also, everything pointed at in giving an ostensive definition is present; and, therefore, the learner would never catch on that it was the present that was being ostensively defined since it could not be contrasted with something which is non-present. In Chapter IV a fuller treatment will be given of the ostensive indefinability of 'time'.[1]

The lesson to be learned from the repeated failures to define 'time' is that this is a misguided approach to the problem of time. As Wittgenstein pointed out, we should not ask what time is, in the sense of asking for a definition or verbal equivalence, but rather should ask how temporal language is used.[2] What is called for is a study of the 'logical grammar', i.e. rules of use, of temporal expressions. To accomplish this and thereby command a clear view of the role of time in our conceptual system we must not only see the logical connections between various types of temporal expressions but also note the way in which these expressions enter into a wide variety of basic concepts. As this investigation proceeds it will become clearer why 'time' is indefinable, this being due to the fact that temporal notions are implicitly involved in all of the basic concepts by means of which we think and talk about the world. It will be seen that temporal notions are like the roots

[1] Augustine's perplexity about time is not due solely to the indefinability of time, but also to his demand that since 'time' is grammatically a noun it function semantically as a name, that is that it name some thing. This is a consequence of his acceptance of a name-substance theory of meaning, the meaning of a word or expression being the object or thing named by it. When it is found that we cannot point to the referent of the noun 'time' it will be felt that time is something mysterious. 'Time' is a name which does not name.

[2] See L. Wittgenstein, *Blue Book* (Oxford, 1960), pp. 6, 26 ff.; and *Philosophical Investigations*, p. 42 ff. The following articles develop Wittgenstein's diagnosis of Augustine's perplexity about time: F. Waismann, 'Analytic-Synthetic, III', *Analysis*, 11 (1952), and also his 'How I See Philosophy', in *Contemporary British Philosophy*, third series, edited by H. D. Lewis (London, 1956), especially pp. 450–453; O. K. Bouwsma, 'The Mystery of Time (Or, The Man Who Did Not Know What Time Is)', *Journal of Philosophy*, LI (1954), reprinted in Bouwsma's *Philosophical Essays* (Lincoln, 1965); R. Suter, 'Augustine on Time with Some Criticisms from Wittgenstein', *Revue internationale de philosophie*, XVI (1962); and R. M. Gale, 'Some Metaphysical Statements about Time', *Journal of Philosophy*, LX (1963).

of a tree which are deeply embedded in the soil of our conceptual system and serve to hold together the different parts of this soil.

The investigation of the role that temporal expressions play in our language of necessity must be selective, for much of ordinary temporal discourse is of no philosophical interest. A philosopher's inspection of the use of ordinary language is always mediated by philosophical problems and perplexities. His concern with ordinary language differs from that of the philologist, grammarian and lexicographer in that he studies those aspects of ordinary language which can help to shed light on, and possibly resolve, traditional philosophical problems. Therefore, it will be wise to begin this investigation by a consideration of the sort of problems and puzzlements that philosophers have raised about time, and then go on to consider the logic of ordinary discourse. This way our attempt to articulate the complex logic of 'time' will have a definite point and direction.

Our investigation will begin by considering McTaggart's famous argument for the unreality of time, which was first published in 1908. The reason for beginning with McTaggart's argument is that it will serve as a key for helping us to understand what twentieth century analytic philosophers of all types have said about time. If one looks carefully enough into the multitudinous writings on time by analysts one can detect a common underlying problem, that being that almost all of them were attempting to answer McTaggart's paradox. This is not to say that all these writings mention McTaggart by name, or even that their authors always had him consciously in mind; but only that the problems they wrestled with were those bequeathed to them by McTaggart. A person can scratch a mosquito bite without knowing that it is a mosquito bite. McTaggart's argument is fallacious, but it is fallacious in such a deep and basic way that an adequate answer to it must supply a rather extensive analysis of the concept of time, along with a host of neighbouring concepts that are themselves of philosophical interest, such as change, substance, event, proposition, truth, etc. What we shall notice in Chapter II is that the answers proposed by analysts involved very different analyses of these concepts. Once we have considered McTaggart's paradox and the competing answers which it has evoked we shall be in a better position to determine what aspects of ordinary temporal discourse are in need of clarification and just how this clarification will help to resolve philosophical problems and perplexities.

McTaggart's paradox is deeply rooted in two fundamentally different ways in which we conceive of and talk about time. On the one hand, we conceive of time in a dynamic or tensed way, as being the very quintessence of flux and transiency. According to this way of conceiving time, events are represented as being past, present and future, and as continually changing in respect to these tensed determinations. Thus what is now happening ceases to happen and becomes past as future events become present or come to pass, this requiring that past events become more past and future events less future. This process of temporal becoming is often referred to by various metaphors and aphorisms—the gnawing tooth of time, the river of time, time flies, here today and gone tomorrow, gather ye rosebuds while ye may, enjoy yourself it's later than you think. This dynamic concept of time lies at the basis of the temporalistic view of man and existence which is presented in certain religions, philosophies and works of art. This conception of time finds expression in our tensed way of talking.

Yet time, in which all things come to be and pass away, necessarily involves a static structure or order. It is, to use T. S. Eliot's phrase, 'a pattern of timeless moments'. The very same events which are continually changing in respect to their pastness, presentness or futurity are laid out in a permanent order whose generating relation is that of *earlier than*. This is the static or tenseless way of conceiving time, in which the totality of history is viewed in a God-like manner, all events being given at once in a *nunc stans*. Here there is no gnawing tooth of time—no temporal becoming—but only a democratic equality of all times. This way of conceiving time finds expression in our tenseless way of talking, in which temporal relations of precedence and subsequence between events are described by timelessly true or false statements.

The problem is to relate these two radically different ways of conceiving or talking about time to each other. As we shall see, McTaggart's paradox rests on a seeming incompatibility between the dynamic and static conceptions of time. To dispel his paradox, we shall have to inspect carefully the logic of our tensed and tenseless ways of talking, and, in particular, see how these two different ways of speaking are related. In regard to the latter question, it will be the aim of this book to show that the logic of ordinary temporal discourse is such that tenseless expressions are logically dependent upon tensed ones, and also that the fundamental concepts which we employ in talking about the world involve tensed concepts; they could not be expressed except in a tensed language. More specifically, an attempt will be made to show the way in which the tensed distinctions of past, present and future are logically contained in our concepts of precedence and simultaneity, things and events, change, causality, action, intention, knowledge, possibility, truth, propositions and identification. If this programme can be successfully carried through it will show that our dynamic or tensed way of conceiving or talking about time is more basic than our static or tenseless view of time. It will turn out that the concepts of past, present and future are the tree-like roots which hold together this multitude of concepts. A revealing analysis of time must trace this elaborate network of roots and in so doing reveal the manner in which these concepts are interrelated. Thus an adequate *analysis* of time is also *synthetic*, which shows the bogus nature of the dichotomy between analytic and synthetic philosophy. In addition, this shows that linguistic analysis, although it has the therapeutic function of dissolving philosophical perplexities and puzzles, can have a far more constructive and synthetic function.

Having sketched the significance which McTaggart's argument has for an investigation of the language of time, we can now consider his argument. The underlying assumption of this argument is that any concept which is contradictory cannot be true of reality. The aim of this argument is to reveal a contradiction in the concept of time, thereby establishing the unreality of time. McTaggart's argument comprises both a positive and a negative thesis. The positive thesis contains an analysis of the concept of time, which McTaggart holds to be the only correct one. The negative thesis attempts to show that this analysis of the concept of time entails a contradiction. It is important to remember that McTaggart had both a positive and negative thesis; for, as shall be seen in the next chapter, certain philosophers have basically agreed with his positive thesis concerning the correct analysis of the concept of time but have denied his negative thesis that this analysis leads to a contradiction, while other philosophers have met his argument by challenging his positive thesis.

1. *McTaggart's Positive Thesis*. According to McTaggart there

are two different kinds of temporal facts. First, there are facts about temporal relations of precedence and subsequence between events, and, second, there are facts about the pastness, presentness and futurity of these same events. Corresponding to the first type of temporal facts is a series of events, called the '*B*-Series', which runs from earlier to later, its generating relation being *earlier* (*later*) *than*; while corresponding to the second type of temporal facts is a series of events, called the '*A*-Series', which runs from the far past through the near past to the present, and then from the present through the near future to the far future. The *B*-Series is unchanging, since the relation of *earlier than* is permanent. The *A*-Series, on the other hand, is a changing series, since the distinctions of past, present and future change with time: at any two different times there are different *A*-Series, the event(s) which is present in one of them not being present in the other. The relation of *earlier* (*later*) *than* will be referred to as a '*B*-relation', and the position of an event in the *A*-Series—its pastness, presentness or futurity—as an '*A*-determination'. A linguistic expression which designates a *B*-relation will be called a '*B*-expression', while an expression (such as the tense of a verb) which designates an *A*-determination of an event will be called an '*A*-expression'.

Events can never change in their position in the *B*-Series. The death of Plato, for example, cannot through diligence and hard work sneak up on the death of Queen Anne, for if one event is earlier than some other event by so many time-units it is always the case that the one is so many time-units earlier than the other. The only change that an event can undergo is a change in its *A*-determination. The death of Plato was future, became less and less future until it finally realized its moment of glory by becoming present, and is now retreating into the more distant past. Such change has been referred to above as temporal becoming. McTaggart's *B*-Series corresponds to what has been called the static temporal and his *A*-Series to the dynamic temporal.

Having distinguished these two different type of temporal series, McTaggart next addresses himself to two basic questions. First, are both the *A*- and the *B*-Series essential for the reality of time?, and, second, which of these series, if either, is the more fundamental? His answer to the first question is that both series are essential: time necessarily involves both a dynamic and a static aspect. His arguments in support of this are directed only to showing the necessity for the *A*-Series. Probably he thought the necessity for the *B*-Series so obvious as to require no further comment: to conceive of a 'time' which admits of no distinctions between earlier and later times is a conceptual absurdity. While philosophers have rarely doubted the necessity for there to be a *B*-Series if there is to be time, several of them have questioned the necessity of the *A*-Series. They have argued that although we always experience events as forming both a *B*- and an *A*-Series the *A*-determinations of events are delusory or purely psychological, so that only the *B*-Series is objectively real. Against this line of reasoning McTaggart advances two arguments, the first showing that change, which is necessary for the reality of time, requires an *A*-Series, and the second showing that *B*-relations depend upon *A*-determinations.

The assumption of McTaggart's first argument—that time requires change—would be agreed upon by everyone. He proceeds by showing that every possible candidate for the title of change other than the change in an event's *A*-determination is logically unthinkable. He points out that it is logically impossible for any feature of the *B*-Series to change: there can be no increase or decrease in its membership, no change in the qualities of any of its member events, and no change in the *B*-relations between these members. The *B*-Series, while necessary for time, is not sufficient, since it cannot give us change. To get change we must invoke the changes in the *A*-determinations of the events which make up the *B*-Series. For events to change in their *A*-determinations there must be an *A*-Series, or, more accurately, a series of *A*-Series.

McTaggart's second argument for the necessity of the *A*-Series is directed towards establishing that the *B*-Series, which is assumed to be necessary for time, logically presupposes an *A*-Series. The argument for this is that the generating relation of the *B*-Series—*earlier than*—can be defined or analysed in terms of *A*-determinations.

> The term *P* is earlier than the term *Q*, if it is ever past while *Q* is present, or present while *Q* is future.[1]

[1] J. M. E. McTaggart, *The Nature of Existence* (London, 1927), Vol. 2, p. 271. McTaggart's argument for 'The Unreality of Time' first appeared in *Mind*, XVII (1908), and has been reprinted in his *Philosophical Studies* (London, 1934). The presentation of this argument in *The Nature of Existence* contains some important additions in which he tries to counter the objections which Russell and Broad raised against his original 1908 paper. His presentation of the argument in *The Nature of Existence*

Thus, if events bear *B*-relations to each other it follows that they also must have *A*-determinations and, as a consequence, form an *A*-Series. There no more could be a *B*-relation between events having no *A*-determination than there could be a harmonic relation between notes having no absolute pitch. This establishes the sufficiency as well as the necessity of the *A*-Series; for the *B*-Series is reducible to the *A*-Series, since the generating relation of the *B*-Series is definable or analysable in terms of *A*-determinations.[1]

In answer to the second question, McTaggart claims that the *A*-Series is more fundamental than the *B*-Series. The reason is that the *A*-Series cannot be derived from any other type of series, while the *B*-Series can be gotten out of a union of an *A*- with a *C*-Series, i.e. a series whose generating relation is non-temporal, such as the series of integers. To join an *A*-Series with a *C*-Series, such as the series of integers, we first correlate the present event(s) in the *A*-Series with one of the integers, such as 0, and then correlate the past and future events in the *A*-Series with the negative and positive integers respectively. Each event in the *A*-Series, other than the present one(s), thereby is correlated with some specific negative or positive integer. An event *x* is earlier than an event *y* if the integer with which *x* is correlated is less than the one with which *y* is correlated. In this way the *B*-Series is formed. Notice that the *B*-Series is a time-series only because its member events form an *A*-Series; were it not for this it would become a non-temporal *C*-Series.

2. *McTaggart's Negative Thesis.* Having established the necessity of the *A*-Series for the reality of time, McTaggart, in his negative thesis, attempts to demonstrate the contradictory and therefore unreal nature of this Series, thereby establishing the unreality of time. Before presenting the main argument for the contradictory nature of the *A*-Series, a subsidiary argument is given. This

[1] In Chapter VI it will be seen that this also shows that the *A*-Series is more fundamental than the *B*-Series. Also it will turn out that this argument for the fundamentality of the *A*-Series is inconsistent with his other argument, summarized in the next paragraph, for the fundamentality of the *A*-Series based on the alleged fact that the *A*-Series cannot be derived from any other series, while the *B*-Series can be derived from a union of an *A*- and a *C*-Series.

---

has been reprinted in *The Philosophy of Time*, edited by R. M. Gale (New York, 1967). This book also contains an extensive bibliography of the literature on McTaggart's paradox.

argument is not put forth as being conclusive, since it is based on the mooted assumption that *A*-determinations are relations, which translated into the linguistic idiom, means that *A*-expressions, such as 'is past (present, future)', are disguised two-place predicates. According to this way of construing *A*-expressions, a statement of the form, '*M* is past (present, future)', in which '*M*' refers to an event, is really a relational statement having the form, '*M* has R (R_1, R_2) to ____.'[1] What kind of relation could R (R_1, R_2) be? and what sort of expression could be inserted in the blank space? It might be thought that R is the temporal relation of *earlier than* (*simultaneous with*, *later than*) and that an event-expression should be inserted in the blank space, which would produce the statement, '*M* is earlier than (simultaneous with, later than) *N*,' in which '*N*' refers to an event in the same *B*-Series as the one to which *M* belongs.

McTaggart is quick to point out that this will not work because the *A*-determination of an event changes whereas temporal relations between events cannot do so. *M* receives its specific *A*-determination not in virtue of its temporal relation to some other event but because it bears an I-don't-know-what relation to some I-don't-know-what entity *X*, which is not an event in the *B*-Series. *M* changes in its *A*-determination because of a change in its I-don't-know-what relation to this unknown entity *X*. Until we can specify what *X* and this relation are we do not have any idea of what an *A*-expression means. *A*-expressions are mere noises signifying nothing.

It is obvious why this subsidiary argument is not put forth as being conclusive, for all that need be done to meet it is to deny that *A*-determinations are relations. It is not patently wrong to claim that *A*-determinations are qualities of events, or, put in the linguistic idiom, that an *A*-expression is a *monadic* predicate describing the property of an event. McTaggart's main argument is neutral as to whether *A*-determinations are relations or properties, and for this reason is put forth as being conclusive.

The main argument is as follows. Because *A*-determinations are changing relations or properties of events, an event which has

[1] My convention for the use of parentheses is as follows. Any one of the words in parentheses can be substituted for the word preceding the parentheses so as to make a new statement. If the *n*th word of the parentheses should be so substituted then the *n*th word in every parentheses in that sentence or context must also be substituted.

any one of the three *A*-determinations of past, present or future will also have the other two, assuming that there is no first or last event in history. This is contradictory, because if an event has any one of these *A*-determinations it logically cannot have either one of the remaining two determinations. For example, it is contradictory for an event to be both past and present, since these are mutually incompatible. Since the *A*-Series harbours a contradiction it cannot be real, and therefore time is unreal.

A titter is the usual reaction to this argument for it seems so easy to explain away this alleged contradiction. The obvious retort is that no event has more than one *A*-determination *at the same moment of time*, and therefore there is no contradiction. Rather it is the case that an event, such as *M*, has these three *A*-determinations *successively*: for example, *M* has been future, will be past and is now present. But to say this, McTaggart claims, must mean that *M* is future at a moment of past time, is past at a moment of future time, and is present at a moment of present time, respectively.

But this reply will not do, since it involves either a vicious circle or a vicious infinite regress. What we have done is to explain away the contradiction of an event having all three *A*-determinations by claiming that it has these determinations successively *at moments of time* in a second-order time-series. But since this second-order time-series is itself a time-series, its members, which are moments of time, must also form an *A*-Series. This is a result of McTaggart's positive thesis that there cannot be a time-series—a *B*-Series—which is not also an *A*-Series. Therefore, every moment of time in this second-order *A*-Series has all three mutually incompatible *A*-determinations; and this involves a contradiction. What we have done is to explain away the contradiction inherent in the first-order *A*-Series only by postulating a second-order *A*-Series. And this is to reason in a vicious circle, since we must presuppose a second *A*-Series to rid the first *A*-Series of contradiction.

If we should try to remove the contradiction in this second-order *A*-Series, owing to all the moments of time in it having all three *A*-determinations, by saying that these moments of time are *successively* future, present and past at moments of time in some third-order *A*-Series, we are merely transferring this contradiction to the third-order *A*-Series. We could, of course, rid this third-order *A*-Series of contradiction by reference to a fourth-order *A*-Series, and so on *ad infinitum*. Not all infinite regresses are vicious, but this one is. The reason is that at any point in this regress at which we stop we are left with a contradictory *A*-Series. The curse of contradiction pursues us down this infinite regress, like a sort of baton that each *A*-Series passes on to the next.

At this point in the dialectic, the initial titters have given way to thoughtful meditation as it dawns on one that McTaggart's paradox, like the great Zenoian paradoxes, admits of no simple answer. The attempts by analytic philosophers to refute McTaggart will now be considered.

II

THE ANSWERS TO McTAGGART'S PARADOX

THE obvious opening move to make when confronted with an argument for the unreality of time is to appeal to common-sense facts. G. E. Moore pointed out that if time is unreal, then there are no temporal facts: nothing is past, present or future, and nothing is earlier or later than anything else. But, plainly, it is false that there are no temporal facts, for it is a fact that I am *presently* inscribing this sentence and that my breakfast yesterday *preceded* my lunch. Since the conclusion of McTaggart's argument entails a false statement, it follows by the rule of *modus tollens* that his conclusion is false. Therefore, his argument contains either a false premiss or a nonsequitur.[1]

Valuable as this opening move is, it is of course nothing but an opening move, for it fails to locate and correct the source of McTaggart's error. The task of the philosopher is not just to remind us that we *were* born and that our breakfast *precedes* our lunch, but to give a coherent account of what we mean by these expressions. The account which McTaggart gives of these expressions seems to result in the absurd conclusion that time is unreal. What must be done is to replace McTaggart's account of time by a more adequate one that will not involve this absurd result. The serious answers to McTaggart's paradox are those which attempted to do just that.

Analysts have given three different type of serious answers to

[1] Moore, *Some Main Problems of Philosophy* (New York, 1962), Chapter XI.

McTaggart: (1) The *B*-Series alone is sufficient to account for time; (2) The *A*-Series alone is sufficient to account for time, but the concept of the *A*-Series does not contain a contradiction; and (3) Either the *A*- or the *B*-Series alone is sufficient to account for time, but they must not be confused with each other otherwise a contradiction arises. Answer (1), to be called the '*B*-Theory Answer', attacks McTaggart's positive thesis because it denies that the *A*-Series is necessary for the reality of time. Answer (2), to be called the '*A*-Theory Answer', basically agrees with McTaggart's positive thesis that the *A*-Series is both necessary and fundamental, but denies his negative thesis that the concept of the *A*-Series is contradictory. Answer (3), to be called the 'Either-Way-Will-Work Theory Answer', attempts to show that whether we affirm or deny McTaggart's positive thesis we can give an adequate account of time. Each of these answers will now be considered separately, with particular attention to the competing analyses of time contained in the first two.

THE *B*-THEORY ANSWER

This answer to McTaggart, along with the theory of time contained in it, which will be called the '*B*-Theory', has been popular with those analysts whose approach to philosophy has been made from the side of mathematical logic and theoretical physics, which includes most logical atomists, logical positivists and rational reconstructionists. The father of the modern version of the *B*-Theory is Bertrand Russell, and among his followers are: R. B. Braithwaite, C. J. Ducasse, A. Grünbaum, A. J. Ayer, W. V. Quine, N. Goodman, D. C. Williams and J. J. C. Smart. Most of these philosophers would agree to the following four tenets, which together constitute the *B*-Theory of time:

(1) The *A*-Series is reducible to the *B*-Series since *A*-determinations can be analysed in terms of *B*-relations between events;

(2) temporal becoming is psychological since *A*-determinations involve a *B*-relation to a perceiver;

(3) the *B*-Series is objective, all events being equally real; and

(4) change is analysable solely in terms of *B*-relations between qualitatively different states of a single thing.

We shall consider each of these tenets separately and then show how collectively they contain an answer to McTaggart's argument.

(1) *The A-Series is reducible to the B-Series since A-determinations can be analysed in terms of B-relations between events.* The main contention of the *B*-Theory of time is that events are not past, present and future simpliciter, but are merely earlier than, simultaneous with[1] and later than other events. Another way of stating this position is that 'is past', 'is present' and 'is future', grammatical appearances to the contrary, are really tenseless two-place predicates that take event-expressions as their arguments. Nelson Goodman has attempted to reduce *A*-determinations to *B*-relations in this manner:

> The 'past', 'present', and 'future' name no times. Rather the 'is past at', the 'is present at', and the 'is future at' are tenseless two-place predicates that may respectively be translated by the tenseless predicates 'is earlier than', 'is at', and 'is later than'.[2]

If such a reduction of *A*-determinations to *B*-relations is sound then the *A*-Series is reducible to the *B*-Series.

An event is not past, present or future simpliciter or intrinsically; it is merely earlier than, simultaneous with or later than some other chosen event. The chosen event is usually either a *linguistic* event consisting in the utterance of a tensed statement about this event in question or a *mental* event consisting of a memory, perception or expectation of it. The first method of reducing *A*-determinations to *B*-relations will be called the 'Linguistic Reduction'; the second will be called the 'Psychological Reduction', and will be considered under tenet (2). The aim of both

[1] We are counting *simultaneous with* as a *B*-relation. Our justification for this is that *simultaneity* (*a*) expresses a permanent temporal relation between events, as does *earlier than*, and (*b*) is needed in the construction of the *B*-Series, since *earlier than* relates classes of simultaneous events.

[2] Goodman, *The Structure of Appearance* (Cambridge, Mass., 1951), p. 295. Goodman's use of 'is at' is synonymous with 'is simultaneous with'. A similar treatment of *A*-determinations is given by the young C. D. Broad in his article, 'Time', in the *Hastings Encyclopaedia of Religion and Ethics* (New York, 1951). This article is a vigorous defence of every tenet of the *B*-Theory. The Hasting's *Encyclopaedia* was first published in 1921. When Broad wrote this article he agreed completely with Russell's theory of time. As we shall see further on in this chapter, Broad reversed himself completely in his subsequent writings on time.

of these reductions is to show that the *A*-Series can be reduced to the *B*-Series and that furthermore the *A*-Series is *subject*ive since *A*-determinations involve a reference to a *subject* as either a language-user or a perceiver. If there were no language-users or perceivers in the world there would still be a *B*-Series; however, there would be no *A*-Series. We shall now take up the Linguistic Reduction.

The first step in this reduction is to point out that *A-statements*, i.e. statements asserting that an event has a given *A*-determination, are situational in that they reveal the speaker's temporal relation to the event reported by his statement. To say that an event occurred or will occur simply means that it occupies a time earlier or later than my statement about it. The situational nature of *A*-expressions, such as tenses, can be made more precise by utilizing Hans Reichenbach's theory of token-reflexivity.[1] By a *token* Reichenbach means an individual sign, for example ink marks on a certain piece of paper or the occurrence of a pattern of noises on some occasion. A word or sentence is a class of similar tokens, a token-reflexive word or sentence being one every token of which refers to itself. In accordance with this account of token-reflexivity, the following analysis, to be called 'TRI', can be given of *A*-statements: the *A*-statement, '*M* is past (present, future)', can be analysed into the statement, '*M is* earlier than (simultaneous with, later than) this token', in which 'this token' refers to the occurrence of the entire sentence token within which it occurs.[2] The philosophical significance of the TRI analysis of *A*-statements, according to Smart, is that it brings out the hidden anthropocentricity or linguocentricity of *A*-expressions because they involve a reference to a token produced by some person.[3]

While the TRI analysis may have the metaphysically important consequence of showing that events have *A*-determinations only in relation to a token event, there is room for doubt whether it has succeeded in reducing *A*-determinations to *B*-relations. The

[1] Reichenbach, *Elements of Symbolic Logic* (New York, 1947), pp. 284–287.

[2] Throughout this book we shall adopt the convention of italicizing tenseless verbs and copulae. The copula 'is' in the *analysans* has been italicized to indicate that it is in the tenseless or timeless present. The reader is reminded of the convention for the use of parentheses contained in footnote 1 on page 12.

[3] Smart, '"Tensed Statements": A Comment', *Philosophical Quarterly*, 12 (1962), p. 264. The same account is given by Smart in his *Philosophy and Scientific Realism* (London, 1963), p. 134.

reason is that the statement in its *analysans*, although it eliminates all tensed verbs and copulae and merely describes a *B*-relation between event *M* and the occurrence of the sentence token, involves the use of a sentence which is subject to exactly the same temporal restrictions in its use for making true statements as is the sentence employed in making the *A*-statement in the *analysandum*: both '*M* is past (present, future)' and '*M is* earlier than (simultaneous with, later than) this token' can be used—uttered or inscribed—so as to make a true statement only if used later than (simultaneous with, earlier than) the occurrence of *M*. Those who have attempted to reduce *A*-determinations to *B*-relations sought to create an *ideal* language. One of the requirements of an ideal language is that the sentences which occur in it be freely repeatable, their use at any time making a statement having the same truth-value. The sentence in the *analysans* of TRI did not achieve this desideratum because it contained the token-reflexive expression 'this token'. To achieve the goal of free repeatability an *A*-statement must be analysed into a 'genuine' *B-statement*, i.e. a statement made through the use of a freely repeatable sentence which describes a *B*-relation between two events.[1]

The second step in the Linguistic Reduction, taken by those *B*-Theorists who seek to create an ideal language, is to go beyond the TRI analysis by eliminating all token-reflexive expressions from the *analysans*, thereby achieving the ideal of free repeatability. The following analysis, to be called 'TRII', analyses '*M* is past (present, future)' into '*M is* earlier than (simultaneous with, later than) *theta*', in which '*theta*' is a metalinguistic proper name for the occurrence of the tensed sentence token in the *analysandum*.[2]

[1] Whether the *analysans* of TRI is an *A*- or *B*-statement cannot be determined until the concepts of an *A*-determination and a *B*-relation are clarified, as they will be in the next chapter.

[2] The TRII analysis is also given by Reichenbach, *op. cit.* Many authors who have defended the token-reflexive analysis of tense have not made it sufficiently clear whether they were advocating TRI or TRII. The following quotations contain such an ambiguity: '"Washington cross*ed* the Delaware" thus means, not merely that the world sum is embellished by Washington's crossing the Delaware, but that this embellishment occurs in that part of the total which lies to the pastward of the temporal meridian defined by the occurrence of this particular token-sentence, and correspondingly for other tenses.' D. C. Williams, 'The Sea Fight Tomorrow', in *Structure, Method and Meaning*, edited by P. Henle, H. Kallen and S. Langer (New York, 1951), p. 287; 'The present tense indicates that the event and the utterance of the sentence recording the event, are nearly simultaneous; the past tense indicates that the record is posterior to the event.' B. Mayo, 'Events and Language', in

Instead of the metalinguistic proper name '*theta*' we could have used a metalinguistic definite description of the token event in the *analysandum*. By this technique '*M* is past (present, future)' would be analysed into '*M is* earlier than (simultaneous with, later than) the occurrence of a token of "*M* is past (present, future)".' This latter statement is freely repeatable, since it mentions rather than uses the tensed sentence '*M* is past (present, future)'. Thus the statement, like the one in the *analysans* of TRII, does not assert that either *M* or the linguistic event consisting in the utterance of the sentence '*M* is past (present, future)' is now past (present, future); it merely describes an unchanging *B*-relation between *M* and the tensed utterance about *M*.

A more common technique for translating an *A*-statement into a *B*-statement is to eliminate in the latter all reference to the occurrence of the original *A*-statement and instead to tenselessly ascribe a date to the event reported by this *A*-statement.[1] For example, 'It is now raining' is rendered by 'Rain *is* at December 30, 1966'. When we tenselessly ascribe a date to an event we show a *B*-relation between this event and the event which serves as the origin of our calendar. In mathematical logic a date is one among sundry predicates that are tenselessly ascribed to something.[2] The characterizing copula as well as the quantifier is tenseless. When a logician says, 'There *is* a sea battle, which *is* at December 30, 1966', he is not saying that either the sea battle or the date in

[1] According to Russell in *Principles of Mathematics* (Cambridge, 1903), p. 471 and his review of H. MacColl's *Symbolic Logic and Its Application* in *Mind*, XV (1906), pp. 256–257, a tensed sentence of the form 'S is now ϕ' is only a propositional function, being of the form 'S *is* ϕ at t', in which 't' is a free variable. The propositional function becomes a proposition, and therefore is true or false, only when a value—a date—is supplied for the variable 't'.

[2] For a defence of this see: Quine, 'Mr. Strawson on Logical Theory', *Mind*, LXII (1953), p. 443; L. J. Cohen's review of A. N. Prior's *Time and Modality* in the *Philosophical Quarterly*, 8 (1958), pp. 267–268; and C. Lejewski's review of Prior's *Time and Modality* in *Philosophy*, XXXIV (1959).

Philosophy and Analysis, edited by M. Macdonald (Oxford, 1954), p. 177. The firs quotation does not tell us how the temporal meridian at which the token-sentence occurs is defined: is it through the use of a token-reflexive word, as in TRI, or through the use of a metalinguistic proper name or definite description, as in TRII? The second quotation is also ambiguous as to just how the relation between event and utterance is indicated—through TRI or TRII. These examples could be multiplied. Usually we have to guess which version of the token-reflexive theory an author is espousing by the inferences he draws. If he draws the inference that the *analysans* is freely repeatable, he is probably accepting TRII rather than TRI.

question is now present; for both the 'is' of the existential quantifier ('There *is* a') and the 'is' of the characterizing copula are tenseless. Quine has claimed that the tenseless mode of existential quantification fits in well with the tenseless space-time talk of Minkowskian geometry employed in relativity theory. In relativity theory, supposedly, a thing is represented as a four-dimensional worm consisting of three-dimensional cross-sections strung along the fourth dimension of time. The physicist speaks in a tenseless mode about these cross-sections, saying, for example, that a certain cross-section *is* earlier or later than some other cross-section.[1] The assumption here is that what is good for physics is also good for logic and metaphysics.

Regardless of which technique we use for analysing *A*-statements into *B*-statements, we are left with 'genuine' *B*-statements—statements that are true once and for all. This holds even if the original *A*-statement was about a contingent event in the future. The fact that language can be 'detensed' without loss of meaning shows that *A*-determinations are not intrinsic to events.

(2) *Temporal becoming is psychological since A-determinations involve a B-relation to a perceiver.* The Psychological Reduction can be viewed as a supplement to the Linguistic Reduction. To the claim that *A*-determinations indicate a *B*-relation between the reported events and utterances the Psychological Reduction adds the claim that *A*-determinations also *express* the characteristic state of belief of the speaker—whether he remembers, perceives or anticipates the reported event.[2] Russell has argued that *A*-determinations are notions derived from psychology, since to understand them reference must be made to consciousness. To understand what is meant by 'past' we must make one of our past experiences an object of experience, while to understand the meaning of 'present' we must refer to one of our sensations, since 'to be present' means to be the object of a sensation.[3] Thus, in a world

[1] Quine, *op. cit.* Also to be consulted is Smart, *Philosophy and Scientific Realism*, Chapter VII.

[2] See Russell, *The Analysis of Mind* (London, 1921), Chapter XII.

[3] Russell, 'On the Experience of Time', *Monist*, 25 (1915), pp. 214–215. Similar views are expressed by: R. M. Blake, 'On Mr. Broad's Theory of Time', *Mind*, XXXIV (1925); R. B. Braithwaite, 'Time and Change', *Aristotelian Society Supp. Vol.*, VIII (1928); C. J. Ducasse, *Nature, Mind, and Death* (LaSalle: 1951), p. 316 ff.; and A. Grünbaum, *Philosophical Problems of Space and Time* (New York, 1963), Chapter 10; and in his article 'The Status of Temporal Becoming', in *The Philosophy of Time*, edited by R. M. Gale (New York, 1967).

devoid of sentient organisms there no more would be *A*-determinations than there would be pains. And since *A*-determinations are psychological it follows that temporal becoming is also. The temporal becoming of a physical event is analysable in terms of its bearing different *B*-relations to a series of mental events, consisting of an expectation, perception and memory, experienced by a single mind.

(3) *The B-Series is objective, all events being equally real.* What is objectively real is the *B*-Series, for *earlier than* does not involve any reference to a *subject* either as a languager-user or perceiver. The *A*-determinations of events were seen to be *subject*ive in the sense of involving a *B*-relation to a linguistic or mental event. There could be *B*-relations between events in a world devoid of language-users or perceivers, but there could not be *A*-determinations and therefore no temporal becoming. Corresponding to our timelessly true *B*-statements there is a timeless series of events—the *B*-Series. According to some *B*-Theorists, all events of history form a *totum simul.*[1] Past, present and future are equally real and determinate. This is a necessity of logic, since we can formulate timelessly true *B*-statements about the *B*-relations between events.

(4) *Change is analysable solely in terms of B-relations between qualitatively different states of a single thing.* It is claimed that the *B*-Series alone is sufficient to account for (i) changes *of* time, i.e. changes in the *A*-determinations of events, and (ii) changes *in* time, i.e. qualitative and quantitative changes of things. The first type of change is temporal becoming, and we have already seen how it is analysable in terms of *B*-relations between events, in which one of the relata is either a linguistic event (the Linguistic Reduction) or a mental event (the Psychological Reduction).

Qualitative and quantitative change—change *in* time—is analysed by Russell solely in terms of the *B*-Series as follows:

> Change is the difference, in respect of truth or falsehood, between a proposition concerning an entity and a time *T* and a proposition concerning the same entity and another time *T′*, provided that the two propositions differ only by the fact that *T* occurs in the one where *T′* occurs in the other.[2]

[1] See Blake, *op. cit.*; Broad, *op. cit.*; and Williams, *op. cit.*, and Williams, 'The Myth of Passage', *Journal of Philosophy*, XLVIII (1951).

[2] Russell, *Principles of Mathematics*, p. 469.

In expounding this position, Braithwaite said that change *in* time 'consists in the fact that a different relation holds between the thing (a substance in the ordinary sense) and a certain property at one of the times than holds at the other time'.[1] An empirical thing or substance is reducible to a certain series of successive events, such that the members of any one such series are intimately interconnected by certain spatial and causal relations which do not interconnect members of any two such series. What we mean by the change of a thing *in* time is a sequence of *successive* events which are all regarded as states of this one thing. Such changes are analysable solely in terms of *B*-relations between events comprising the history of a single thing without invoking the temporal becoming of these events. Time is nothing but a *B*-Series—a certain ordered extension of events.

Answer to McTaggart's Paradox. McTaggart, in arguing for the necessity of the *A*-Series, claimed that change, without which there could not be time, required the *A*-Series. It is logically impossible for any feature of the *B*-Series to change: there can be no increase or decrease in its membership, no change in the qualities of any event included in it, and no change in the *B*-relations between its members. Therefore, by a process of elimination, the only possible change is change in the *A*-determination of an event. The *B*-Theory claims that McTaggart considered only changes in *events* and thus overlooked one possible candidate for the title of change, namely changes in *things*. And in tenet (4) of the *B*-Theory it was argued that such changes in things—changes *in* time—require only the *B*-Series. Therefore, the reality of time does not require that there be an *A*-Series.

While the *B*-Theory denies the necessity of the *A*-Series, it still cannot countenance McTaggart's claim that the *A*-Series is contradictory; for in tenet (1) it claims that the *A*-Series is reducible to the *B*-Series. Such a reduction would not be possible if the *A*-Series really harboured a contradiction. Therefore it behooves *B*-Theorists to answer McTaggart's main argument to the effect that the *A*-Series is contradictory because every one of its members has all three mutually incompatible *A*-determinations. Their way of explaining away this alleged contradiction is to claim, in accordance with their Linguistic and Psychological Reductions, that the *A*-determinations of past, present and future

[1] Braithwaite, *op. cit.*, p. 169.

are really disguised *B*-relations, meaning respectively earlier than, simultaneous with and later than. There is no more a contradiction in an event being past, present and future than there is in an event being earlier than, simultaneous with and later than. A contradiction would emerge only if an event had two or more of these temporal relations to one and the same event; but this McTaggart has never shown to be the case. By the Linguistic Reduction it is shown that an event is past, present and future only in the elliptical sense of being respectively earlier than a tensed utterance that it did occur, simultaneous with an utterance that it is occurring and later than an utterance that it will occur.[1] And by the Psychological Reduction it is claimed that an event has these three *A*-determinations only in virtue of its being respectively earlier than a memory of it, simultaneous with a perception of it and later than an expectation of it.

THE *A*-THEORY ANSWER

This answer to McTaggart is based on what we shall call the '*A*-Theory of time'. The *A*-Theory of time received its most articulate formulation in the work of C. D. Broad, and it has been prevalent among those philosophers whose main concern is to clarify ordinary language by studying the way words are used in various contexts. Among those holding views in some ways similar to Broad's are: L. S. Stebbing, John Wisdom, P. Marhenke, A. N. Prior, D. Pears, W. Sellars, S. Hampshire, P. F. Strawson and J. N. Findlay. These men have far less in common than the defenders of the *B*-Theory, making it more difficult to abstract a set of common tenets from their writings; but I believe that the following four tenets, which together constitute the *A*-Theory of time, are representative of their work:

(1) The *B*-Series is reducible to the *A*-Series since *B*-relations can be analysed in terms of *A*-determinations;
(2) temporal becoming is intrinsic to all events;
(3) there are important ontological differences between the past and the future; and
(4) change requires the *A*-Series.

[1] This way of refuting McTaggart is developed by Broad, *op. cit.*; Braithwaite, *op. cit.*; and D. W. Gotshalk, 'McTaggart on Time', *Mind*, XXXIX (1930), p. 33.

Each one of these tenets is either the contradictory or the contrary of the *B*-Theory tenet of the same number. As before, we shall discuss each tenet separately and then show how together they constitute an answer to McTaggart.

(1) *The B-Series is reducible to the A-Series since B-relations can be analysed in terms of A-determinations.* The defenders of the *A*-Theory agree with McTaggart's positive thesis that the *A*-Series is both necessary and fundamental. We shall first consider their arguments for the necessity of the *A*-Series, which attempt to show that *B*-relations are *temporal* relations only because their relata have *A*-determinations and change in respect to them.

One argument states that what makes the *B*-Series a temporal series is that its members form an *A*-Series and change in respect to their *A*-determinations. Broad, in arguing for this, made two claims, one highly controversial and the other acceptable even to those holding the *B*-Theory of time.[1] The non-controversial claim is that a temporal series differs from a spatial one in that, while both have an intrinsic order, only the former has an intrinsic direction or sense. The controversial claim is that the direction of a temporal series is due to the fact that the members of this series undergo temporal becoming. First we shall explain what it means to say that a temporal series has an intrinsic direction as well as an intrinsic order, and shall then show why Broad believes that this direction can be accounted for only by the temporal becoming of the members of this series.

A linear spatial order of one dimension has an intrinsic order, because, given any three members of this order, one of them will appear to be between the other two regardless of what position they are viewed from. This order, however, does not have an intrinsic direction because whether one member of it is to the right of another will depend on the position from which they are viewed. An observer in the orchestra will see Warren standing to the right of Tebaldi, but an observer backstage will see Warren to the left of Tebaldi. A linear spatial order has a direction only in reference to the right and left hands of an *external* observer; thus

[1] Broad, *Examination of McTaggart's Philosophy* (Cambridge, 1938), Vol. II, p. 269. Broad's section on time in this volume is reprinted in *The Philosophy of Time*, edited by Gale (New York, 1967).

its direction is extrinsic to the order itself. The spatial relation *to the right of* is triadic, since it involves a relation to some third position from which one object is to the right of another. In a linear spatial series there is no asymmetric dyadic relation intrinsic to the series.

Like a spatial series of one dimension, a temporal series has an intrinsic order; but, in addition, it has an intrinsic direction, since its generating relation—*earlier than*—is a genuine dyadic relation. Max Black, who has made an analysis of temporal relations similar to Broad's, states that temporal relations, unlike spatial ones, are not *incomplete relations*, since they do not involve reference to some third term.[1] If we say that *X* is earlier than *Y* we do not have to specify some third position from which *X* and *Y* are viewed, as we do when we say that *X* is to the right of *Y*. When we are told that Warren's singing of a rousing aria was earlier than his dropping dead we do not ask from what third position these events were viewed. Broad claims that temporal betweenness is not fundamental (as is spatial betweenness) because it is definable in terms of the genuinely dyadic relation of *earlier than* taken twice over.

The controversial claim made by Broad is that becoming, what he calls the 'transitory aspect of temporal facts', is of the very essence of time, and that without it a temporal series would not have an intrinsic direction, thereby being indistinguishable from a spatial order. It is because and only because the events comprising a *B*-Series undergo temporal becoming that we can distinguish a *B*-Series from non-temporal series. We cannot do this solely in terms of the logical properties of the generating relation of the *B*-Series—its being asymmetric, transitive and irreflexive—since there are non-temporal relations, such as *larger than*, which have the same logical properties: and yet we know that a number series is basically different from a time-series. It is temporal becoming that causes all analogies between space and time to break down. When we represent time by a line we cannot account for the direction of a time-series. If direction is to be introduced into a one-dimensional spatial order this must be done extrinsically, either by reference to motion along the line (which therefore includes time since motion takes time), or by reference to the right and left hands of an external observer. The intrinsic direction of

[1] Black, 'The "Direction" of Time', *Analysis*, 19 (1959), p. 60.

the *B*-Series is therefore essentially bound up with the temporal becoming of its members.[1]

This argument of Broad's, if it can be called such, will not convince any *B*-Theorist. He would agree with Broad that *earlier than* is a dyadic relation and that a time-series is *sui generis*, but would deny that these two facts are dependent upon temporal becoming. What makes a time-series unique, he would contend, is that its generating relation is a *temporal* relation. Certain empirically minded *B*-Theorists, such as Adolf Grünbaum, would argue that what bestows a direction on a time-series, what makes it irreversible or anisotropic, are certain physical asymmetries, such as increase of entropy, which are intrinsic to the *B*-Series of physical events. The efforts of such theorists to explicate the direction of time make no reference to temporal becoming, being concerned only with law-like and/or boundary condition asymmetries between earlier and later times.

The second argument presented by Broad and other proponents of the *A*-Theory of time for the necessity of the *A*-Series is that *B*-relations are definable or analysable in terms of *A*-determinations. In addition to showing the necessity for the *A*-Series, this also is an argument for the fundamentality of the *A*-Series, since it attempts to show that the *B*-Series is reducible to the *A*-Series. This argument claims that '*M is* earlier than *N*' means 'When *M* is present *N* is future and when *N* is present *M* is past.' This is similar to McTaggart's definition of *B*-expressions in terms of *A*-expressions, which is given on page 10.[2]

Since *B*-relations are definable or analysable in terms of *A*-determinations it follows that *A*-statements cannot be translated without loss of meaning into 'genuine' *B*-statements. The main difficulty with all such translations is that the *B*-statement does not assert, as does the *A*-statement of which it is a rendering, whether the reported event is past, present or future. There is an asymmetry in information between *A*- and *B*-statements. When I tell you that *X* is present and *Y* is future I have already told you that *X*

[1] See Broad, *Scientific Thought* (London, 1925), pp. 57–58; and Black's review of Whitrow's *The Natural Philosophy of Time* in *Scientific American*, V. 206 (1962), p. 183.

[2] Attempts to analyse *B*-relations in terms of *A*-determinations are found in: Broad, *op. cit.*, p. 303; Deshpande, 'Professor Ayer on the Past', *Mind*, LXV (1956), p. 86; and W. Sellars, 'Time and the World Order', in *University of Minnesota Studies in the Philosophy of Science*, edited by G. Maxwell, Vol. III (Minneapolis, 1957), p. 546, reprinted in Sellars, *Science, Perception, and Reality* (New York, 1963).

is earlier than *Y*; but if instead I had asserted the latter *B*-statement, I would not have told you that *X* is present and *Y* is future. Attempts to detense language by the use of dates fail for the same reason.[1]

We are not able to assimilate tensed existential statements to the tenseless mould of existential quantification, as the *B*-Theory demands. Ordinary language, as Strawson and Hampshire have argued,[2] is an irreducibly tensed-substance language because of necessity it must contain referring expressions, and only tensed statements can refer successfully to unique objects and events. In order to single out and describe some object in our immediate environment we must speak in a tensed mode, using ourselves as centres of spatio-temporal orientation. When we speak about objects (or events) in ordinary contexts the question of whether they exist (or occur) now is always relevant. For this reason the method of existential quantification employed in mathematical logic cannot do justice to ordinary language.[3]

Broad attacks the first step in the Linguistic Reduction by denying that *A*-statements are token-reflexive. Even if it were possible for these statements to be token-reflexive, it would still not eliminate a reference to the *A*-determinations of events, for supposedly the tensed utterance-token I refer to when I make a tensed statement differs from all other utterances I make in that it is present while they are past or future. In other words, if '*M* is present' means '*M is* simultaneous with this token' there is still a reference to the *A*-determination of an event, for by 'this token' I mean 'the token which I am uttering or inscribing at present'. If my tensed utterance-tokens are intrinsically past, present and future, how is it possible for events bearing *B*-relations to these tokens not also to be intrinsically past, present and future?[4]

(2) *Temporal becoming is intrinsic to all events.* A similar argument can be employed against the Psychological Reduction, which said that events have *A*-determinations only in relation to a perceiver.

[1] See Prior, 'Thank Goodness That's Over', *Philosophy*, XXXIV (1959), p. 17; and L. E. Palmieri, 'Empiricism and a Time-Line', *Philosophical Quarterly*, 10 (1960), p. 166.

[2] Hampshire, *Thought and Action* (New York, 1960), Chapter One; Strawson, *Individuals* (London, 1959), Part I.

[3] Strawson, *Introduction to Logical Theory* (London, 1952), p. 151. Cf. Prior, *Time and Modality* (Oxford, 1957), p. 26.

[4] Broad, 'Time and Change', *Aristotelian Soc. Supp. Vol.* VIII (1928), p. 187.

But once again this does not reduce *A*-determinations to *B*-relations, because the series of mental events that runs concurrently with the series of physical events forms an *A*-Series. I am sure that my own mental events temporally become. If the mental acts of an observer are intrinsically past, present and future, then the objects of these acts must likewise be intrinsically past, present and future. Becoming cannot be intrinsic to one series of events without being intrinsic to the other.[1]

(3) *There are important ontological differences between the past and the future*. The *A*-Series is objective, not being dependent upon a language-user or perceiver. Several *A*-Theorists have gone on to claim that because temporal becoming is objective there are significant ontological differences between past and future events, consisting in the fact that the future is open, a realm of possibilities, while the past is closed, a realm of actualities. This ontological difference reveals itself in certain logical asymmetries in our ways of talking about the past and the future in that (i) all statements about the past are either true or false while some statements about the future are neither true nor false now, and (ii) statements about the future must be general in logical form while statements about the past can be singular. To say that a statement is singular means that its subject term identifies a particular individual. These two purported logical asymmetries are independent of each other; one can, for example, deny (i) by holding that all statements about the future are either true or false, and, without inconsistency, deny that any statement about the future can be singular. Because a statement is general, e.g. 'Someone will be born tomorrow', does not debar it from being true or false now, and conversely a singular statement, such as 'Joe Smith will die fifty years from now in a plane crash', might be neither true nor false now.

The *B*-Theory cannot countenance these two asymmetries because it allows for no ontological distinctions between events occurring at different times. All events comprising history are equally real and determinate, and can be completely described by timelessly true or false *B*-statements. By a parity of reasoning, it should also be true that all individuals, regardless of when they

[1] This is defended by Marhenke, 'McTaggart's Analysis of Time', in *The Problem of Time*, California University Publications in Philosophy, Vol. 18 (1935), p. 162; G. J. Whitrow, *The Natural Philosophy of Time* (London, 1961), Chapter VI; and M. Capek, *The Philosophical Impact of Contemporary Physics* (Princeton, 1961), Chapter XVII.

exist, are equally real and determinate, and therefore should be identifiable through the use of suitable referring expressions, such as proper names and definite descriptions.

(4) *Change requires the A-Series.* The *B*-Theory analysed change *in* time in terms of *B*-relations between events or states which belong to the history of a single thing. The *A*-Theory challenges this analysis, for in tenet (1) it claims that *B*-relations are dependent upon *A*-determinations. From this it follows that since (*a*) change *in* time necessarily involves *B*-relations between events and (*b*) *B*-relations depend upon the *A*-determinations of the events related, there can be no change *in* time without the *A*-Series. Thus the claim that there can be no change in time without the *A*-Series is a corollary of tenet (1). Only if *B*-relations were independent of *A*-determinations could change in time be explained *solely* in terms of *B*-relations between events.

Broad claims that change *of* time—temporal becoming—is *sui generis* and that any attempt to represent it in terms of change *in* time is vicious.[1] Though a thing exists both before and after it undergoes some qualitative or quantitative change, an event does not exist both before and after it becomes present. When we represent becoming by analogy with motion along a one-dimensional spatial order we are reducing change of time to change in time. We think of the present as a spotlight that plays along a line of chorus-girl-like-events; but this involves us in the pseudo-problem of how fast—how many seconds per second—this spotlight moves. All representations of becoming in terms of some kind of motion (or other change in time) involve a vicious circle, since motion (or any other change in time) presupposes becoming.

Answer to McTaggart. The *A*-Theory agrees with McTaggart's positive thesis concerning the necessity and fundamentality of the *A*-Series, but disagrees with his negative doctrine that the *A*-Series involves a contradiction. McTaggart's inability to combine the static view of time exemplified in his *B*-Series with the dynamic view of time contained in his *A*-Series is due to his substantialization of the events in the *B*-Series, which thus causes him to represent temporal becoming as if it were some sort of motion, the *A*-Series moving up the *B*-Series. McTaggart substantialized events

[1] Broad, *Examination of McTaggart's Philosophy*, *op. cit.*, p. 281, and Broad's 'Reply to My Critics', in *The Philosophy of C. D. Broad*, edited by P. Schillp (New York, 1960), pp. 766–767.

because he thought—mistakenly—that the events in the *B*-Series must always co-exist since *B*-relations are permanent. The events in the *B*-Series are now treated like a line of chorus girls, who exist both before and after they do their brief bit in the spotlight of presentness. But that view involves us in the above difficulties of reducing temporal becoming to movement. Broad claims that there is no incompatibility between temporal becoming and the 'permanency' of *B*-relations between events.[1]

The *B*-Theory explained away the contradiction inherent in each event having all three *A*-determinations by showing that these determinations are reducible to *B*-relations. The *A*-Theory cannot avail itself of this way out, since it denies that such a reduction is possible. Rather it claims that since temporal becoming is intrinsic to all events there is no contradiction to be explained away. A contradiction would arise only if in a single utterance an event was asserted to have two or more *A*-determinations, but there is nothing in our temporal experiences to ever warrant making such a statement.[2] *A*-determinations are unanalysable. '*M* is present', according to Stebbing and Wisdom,[3] entails, but does not mean, '*M* is present at a moment which is present'. McTaggart's infinite regress it follows is benign; it is a regress of entailings and not meanings. Therefore, there is no need to analyse statements predicating an *A*-determination of an event in terms of a moment in some second-order *A*-Series at which the event has this *A*-determination.[4]

THE EITHER-WAY-WILL-WORK THEORY ANSWER

The third answer to McTaggart, which is put forth by J. N. Findlay and J. J. C. Smart, is based on a theory of time that claims that either the *A*- or the *B*-Series alone is a sufficient account of time and change, and contradiction arises only if they are confused with each other. The essential tenet of this theory is that the

[1] The same reply to McTaggart is made by Stebbing, 'Some Ambiguities in Discussions Concerning Time', in *Philosophy and History*, edited by Klibansky and Paton (Oxford, 1936), p. 111; and M. Cleugh, *Time* (London, 1937), pp. 150–151.

[2] This is a paraphrase of Broad, *Examination of McTaggart's Philosophy*, *op. cit.*, p. 277. Cf. E. Hall, 'Time and Causality', *Philosophical Review*, XLIII (1934), p. 339.

[3] Wisdom, 'Time, Fact and Substance', *Proceedings of the Aristotelian Society*, XXIX (1928–29), p. 83; and Stebbing, *op. cit.*, p. 117.

[4] Cf. D. Pears, 'Time, Truth and Inference', in *Essays in Conceptual Analysis*, edited by A. Flew (London, 1956), p. 234.

tensed and tenseless modes of speaking are both equally workable and legitimate: everything sayable in a language containing only *A*-statements is equally well sayable in a language containing only *B*-statements, and vice versa. Herein the Either-Way-Will-Work Theory is closer to the *B*-Theory than the *A*-Theory, since one of the essential tenets of the *A*-Theory is a denial that *A*-statements can be rendered without loss of meaning by *B*-statements.

While the tensed and tenseless modes of speaking are both equally workable and legitimate, Findlay points out that there are certain practical advantages in speaking tenselessly since our statements then are timelessly true or false independently of when they are made. The ideal of science is to eliminate tenses so that statements can be freely repeatable. In the tensed mode of speaking we must systematically alter the tense of our statement depending upon the *B*-relation between our utterance and the reported event. McTaggart's paradox results from the attempt to speak at one and the same time in both the tensed and tenseless modes. In certain 'moods' we become enamoured with the time-independence of our tenseless mode of speaking, and we attempt to apply this ideal to the tensed mode: we then are apt to think that there is something contradictory about tenses since we must continually change the tense of our statements with the passage of time. When McTaggart claims that the *A*-Series is contradictory because every event in it has all three *A*-determinations, he is applying the ideal of the tenseless mode of speaking to the tensed mode. A contradiction results only if we expect tensed statements to have the same logical behaviour as tenseless ones, namely to be freely repeatable.[1]

Smart defends a similar thesis. He distinguishes between an objectionable and an unobjectionable sense of spatializing time. In the objectionable sense we conceive of events as being in time in exactly the same way that ordinary empirical substances are in space. We must conceive of them as co-existing substances which endure in some meta-time. In the unobjectionable sense of spatializing time we treat time as a fourth dimension, as in the four-

[1] Findlay, 'Time: A Treatment of Some Puzzles', in *Logic and Language*, Series One, edited by A. Flew (Oxford, 1951), pp. 52–54. In this article Findlay is practising therapeutic linguistic analysis of the later Wittgenstein. In his later article, 'An Examination of Tenses', in *Contemporary British Philosophy*, third series, edited by H. Lewis (London, 1956), he takes a more speculative approach to the problem of time and arrives at a position which is close to the *A*-Theory of time.

dimensional space-time geometry of Minkowski. Herein time is a dimension in which the three-dimensional spatial cross-sections are strung out in accordance with the *B*-relations between them. A system of solid geometry, regardless of how many dimensions it deals with, has no reference to change or time. In a four-dimensional geometry we do not think of space or the network of spatial relations as enduring or lasting throughout some period of time. When we represent time spatially as a network of *B*-relations, we must not apply tensed concepts of enduring or changing to these relations. If we do we are spatializing time in the objectionable sense, since we would then need a meta-time for this change or changelessness to be in. This is the start of the McTaggart infinite regress of times changing or enduring within times.[1]

[1] This is a paraphrase of Smart, 'The River of Time', in *Essays in Conceptual Analysis*, *op. cit.*, p. 223, and his 'Spatialising Time', *Mind*, LXIV (1955), p. 240, reprinted in *The Philosophy of Time*, *op. cit.* Like Findlay, Smart took a more metaphysical approach to the time problem in his later writings, but, unlike Findlay, he adopts the *B*-Theory of time. See footnote 3 on page 18 for a list of these later writings.

Part Two

A-DETERMINATIONS AND *B*-RELATIONS

III

THE CRITERIA FOR *A*- AND *B*-STATEMENTS

OUR historical exposition of the post-McTaggart writings on time by analytic philosophers is now complete. Bearing in mind the problems and questions raised by them, we now have an idea of what aspects of ordinary temporal discourse are of philosophical importance, in that clarification of this discourse will shed light on, and possibly answer, these philosophical problems and questions. At the centre of the dispute between the *A*- and *B*-Theories of time is the relation between tensed and tenseless discourse, more generally between *A*- and *B*-expressions. Accordingly, the remaining three Parts of this book will be devoted to clarifying the logic of these features of ordinary temporal discourse. In the course of doing this, each one of the tenets of the *A*-Theory will be defended against the rival tenet of the *B*-Theory. This Part will argue for tenet (1) of the *A*-Theory, which holds that the *B*-Series is reducible to the *A*-Series because *B*-relations are analysable into *A*-determinations. Since tenet (4) of the *A*-Theory, to the effect that change requires the *A*-Series, is a corollary to tenet (1), this also establishes tenet (4). In Part III a family of interrelated logical asymmetries between the past and the future will be articulated, this giving support to tenet (3) that there are significant ontological differences between the past and the future. In Part IV tenet (2)—that temporal becoming is intrinsic to all events, whether mental or physical—will be defended against both the Linguistic and Psychological Reductions of the *B*-Theory, whose aim is to show the subjectivity of temporal becoming.

That the *B*-Series is reducible to the *A*-Series is to be supported by showing that *B*-relations are analysable in terms of *A*-determinations. The latter in turn will be supported by showing that a *B*-statement can be analysed in terms of *A*-statements. Before attempting to do this, it is necessary to develop an adequate criterion for both an *A*- and a *B*-statement. Without such a criterion there is no way to assess the success of an attempt to analyse *B*-statements into *A*-statements, or vice versa. It was the absence of such a criterion that caused *A*- and *B*-Theorists to argue at cross-purposes: for example, there was no way of deciding (on pages 18–19) whether the TRI analysis of the *B*-Theory successfully reduced *A*-statements to *B*-statements since it was not clear whether the *analysans* contained a *B*-statement, and similarly it was not possible to judge the success of the attempts by McTaggart (on page 10) and various *A*-Theorists (on pages 27–28) to analyse *B*-statements in terms of *A*-statements. If there is to be any fruitful discussion of the problems of time raised by the *B*- and *A*-Theories it is necessary that at the outset a satisfactory criterion be given for *A*- and *B*-statements. We now turn to this vital task.

We shall first attempt to formulate a language-independent criterion for an *A*-statement. A rough and informal definition of an *A*-statement was given on page 19, where it was said that an *A*-statement asserts that an event has a given *A*-determination. It might be thought that this could serve as the desired language-neutral criterion because it makes no mention of any iconic or grammatical feature of sentences that are used to make *A*-determinations, as for example that these sentences must contain certain words, such as 'present', or that they must satisfy some grammatical requirement, such as containing a tensed verb or copula. However, if this criterion is adopted it begs the question at issue between the *A*- and *B*-Theory, for tenet (1) of the latter attempts to analyse *A*-determinations in terms of *B*-relations by showing that an *A*-statement can be translated into a *B*-statement. For this translation to succeed it is necessary for the *B*-statement to convey the same information as the *A*-statement of which it is a translation. This means that the *B*-statement, like the *A*-statement, would have to convey information about an event's *A*-determination, but then, according to this criterion, it would be an

A-statement. And, as a result, the *B*-Theory would have failed to translate an *A*-statement into a *B*-statement. So as not to beg the question, this criterion can serve only as a necessary but not as a sufficient condition for being an *A*-statement.

A useful strategy for determining the criterion for an *A*-statement is to pick out sentences in some language which unquestionably are used to make *A*-statements and then to see what rules of use they obey. This way a language-independent criterion can be discovered, for these rules of use will be inter-linguistic as they do not specify any iconic or grammatical requirement for those sentences which satisfy them, just as analogously the rules for the positioning and movement of the king in chess do not mention what size, shape and material a chessman must be if it is to play the role of the king in a specific game of chess. Being a native speaker of the English language, I shall begin by locating English sentences which unquestionably are used to make *A*-statements. From a study of the use of these paradigmatic sentences a set of inter-linguistic rules will be extrapolated, thus yielding the desired language-independent criterion for an *A*-statement. Armed with this criterion we can determine what sentences in other languages, as well as certain unfamiliar sentences in English, also serve the function of making an *A*-statement.

Examples of such paradigmatic English sentences are sentences which contain: (1) a tensed verb, e.g. 'Jones is now (was, will be) running'; (2) a tensed copula with a predicate referring to some sensible property, e.g. 'The table is now (was, will be) blue'; (3) a tensed existential quantifier with an event expression, e.g. 'There is now (was, will be) a falling of the barometer'; and, more rarely (4) a tensed copula with a temporal adjective, e.g. 'The drought is (was, will be) present (past, future)'. The latter, it will be seen, is not as paradigmatic as the first three. These four techniques for making *A*-determinations are grammatically dissimilar in certain respects, although they all involve the use of a tensed verb or copula. Here is our class of paradigms; by studying the rules of use which they obey, a language-independent criterion for an *A*-statement can be derived.

What sentences of the form (1)–(4) have in common is that there are *temporal* restrictions placed upon their use for making true (false) statements. They can be used—uttered or inscribed—only at certain times so as to make a true statement; their use at other times makes a false statement. This suggests that an *A*-statement is one which can be made only through the use of a sentence which is subject to these temporal restrictions in its use. Therefore, as a first approximation to an adequate criterion for an *A*-statement, it can be said that:

(*A*) Any statement is an *A*-statement if, and only if, it is made through the use of a sentence for which it is logically possible that two non-simultaneous uses of this sentence make statements differing in truth-value.

A sentence that satisfies this criterion will be called an '*A*-sentence'. By a 'sentence' is meant a sentence type, as contrasted with a sentence token: a sentence type is a class of physically similar oral sentence tokens and physically similar inscribed sentence tokens which are the written expressions of these oral tokens. It is apparent that English sentences of type (1)–(4) above satisfy criterion (*A*). Even if Jones, being an unusually energetic sort, is running at every time in history, it is still *logically* possible for two non-simultaneous uses of the sentence 'Jones is now running' to make statements differing in truth-value.

This criterion applies to molecular as well as atomic statements. Because the constituent statements of a molecular statement are *A*-statements does not ensure that the molecular-statement also is an *A*-statement. For example, the disjunction, 'Either there did exist a tiger or there now exists a tiger or there will exist a tiger', contains three disjuncts each of which satisfies (*A*); and yet the entire disjunctive statement fails to satisfy (*A*). There is nothing paradoxical about this, for this disjunction means the same as, 'There *exists* a tiger.' Another consequence of (*A*), which is to its credit, is that it rules out temporally ubiquitous statements of the form '*S* is always ϕing' as *A*-statements, since it is not logically possible for two non-simultaneous uses of the sentence, '*S* is always ϕing', to make statements differing in truth-value. The fact that the statement that *S* is always ϕing entails *A*-statements, such as that *S* is now ϕing, is no reason for classifying it as an *A*-statement.

There are problems with criterion (*A*) that will require it to be restricted in different ways. The first problem has to do with necessarily true (false) statements that are made through the use of tensed atomic or molecular sentences. The analytically true statement, 'This bachelor is now unmarried', is made through the use of a tensed sentence of type (2) above; yet it fails to satisfy (*A*) because it is not logically possible for two non-simultaneous uses of it to make statements differing in truth-value. The same problem arises for necessarily false statements made through the use of a tensed sentence, such as 'This bachelor is now married.' The easiest way out of these difficulties is to restrict (*A*) so that it does not apply to necessarily true (false) statements:

(*A*)^1 Any statement which is not necessarily true (false) is an *A*-statement if, and only if, it is made through the use of a sentence for which it is logically possible that two non-simultaneous uses of this sentence make statements differing in truth-value.

This is not a purely *ad hoc* restriction, since there is, as Wittgenstein pointed out in the *Tractatus* (4.46 ff.), an important sense in which tautologies and contradictions are not statements: they assert or state nothing about the world.

A second difficulty with (*A*), even when amended as (*A*)^1, is that it fails to distinguish between sentences which are not freely repeatable because they contain a temporal indexical sign and those which are not freely repeatable because they contain some non-temporal indexical sign; as a result, this criterion can serve only as a necessary condition for being an *A*-statement. Roughly, an 'indexical word' is one whose referent depends on the speaker's spatio-temporal relation to his environment.[1] To know the referent of a use of an indexical word, such as 'I', 'you', 'here', 'now' and 'this', we must know respectively who uses the word, to whom he speaks, where, when and what he is pointing at. We would not want to classify a statement of identity as an *A*-statement: the copula of such a statement is not in the present tense. Yet the sentence, 'I *am* Richard Gale', satisfies (*A*)^1, since it is logically possible that two non-simultaneous uses of it make statements differing in truth-value. This would happen if one of these statements was made by me and the other by someone else. Another example of a sentence which is non-freely repeatable

[1] I borrow this term from C. S. Peirce. For a more full treatment of indexical words and related topics see my article, 'Indexical Signs, Egocentric Particulars, and Token-Reflexive Words', in *The Encyclopedia of Philosophy*, edited by P. Edwards (New York, 1967).

solely in virtue of its containing a non-temporal indexical word is, 'Here is where *S* is always ϕing.' We have already seen that the temporally ubiquitous statement, '*S* is always ϕing', is not to be classified as an *A*-statement. But the sentence, 'Here is where *S* is always ϕing', logically could be used at different times to make statements differing in truth-value, even if it is true that *S* is always ϕing. This could happen if the place referred to by the spatial indexical word 'here' shifted from one use to another.

The problem of non-temporal indexical words can be handled by restricting $(A)^1$ as follows:

> $(A)^2$ Any statement which is not necessarily true (false) is an *A*-statement if, and only if, it is made through the use of a sentence for which it is logically possible that two non-simultaneous uses of this sentence make statements differing in truth-value, even if both statements refer to the same things and the same places.

I believe that this new criterion will do the trick of ruling out sentences which are non-freely repeatable solely in virtue of their containing a non-temporal index. The sentence, 'I (This man) *am* (*is*) Richard Gale', does not satisfy $(A)^2$. A person is at least a *thing* (object or substance in the ordinary sense). It is not logically possible for two non-simultaneous uses of this sentence to refer to the same *things*—in this case the same person—and yet make statements differing in truth-value. 'Here is where *S* is always ϕing' also fails to satisfy $(A)^2$, since it is not logically possible for two non-simultaneous uses of this sentence to refer to the same place(s) and yet make statements differing in truth-value. Because we require that the same *things* be referred to by two non-simultaneous uses of a sentence does not rule out as an *A*-statement a statement describing a spatial relation between two *things*. For example, the sentence, 'This man is now ten feet from that man', does have the function of making an *A*-statement, because, even when the *things* referred to by 'this' and 'that' are kept constant, it is possible for the spatial relation between them to change and therefore for two non-simultaneous uses of this sentence to make statements differing in truth-value. The requirement that the same *things* be referred to on each occasion when the sentence is used needs further explanation. It is of utmost importance that we do not require that the same *events* or *states of affairs* be referred to by each use of the sentence. The sentence, 'Jones is now winning this race', is used to make an *A*-statement: its demonstrative 'this' refers to a race, and since a race is not a *thing* but an *event*, it is not ruled out by the requirement that the same *things* be referred to by each use of the sentence.

We not only want a language-independent criterion for an *A*-statement but also for different species of *A*-statements, i.e. statements which make determinations of the presentness, pastness or futurity of some event or state of affairs. Once again this can be accomplished by studying the rules of use for the paradigmatic tensed English sentences (1)–(4) above. The rules for the use of these tensed sentences specify that a past (present, future) tense sentence can be used to make a true statement only if used later than (simultaneous with, earlier than) the event or state of affairs reported by the statement which is made. These rules specify the condition under which a tensed sentence can be used to make a true statement, and thus they could appropriately be called 'Truth-condition Rules for the Use of Tensed Sentences for Making True Statements', which will be abbreviated as 'Truth-condition Rules for Tensed Sentences'.[1] These Rules are language-neutral because they make no reference to vocabulary or grammar. Therefore, they can be generalized to cover all *A*-sentences and can be called 'Truth-condition Rules for *A*-sentences'. These Truth-condition Rules for *A*-sentences can be utilized to formulate language-neutral criteria for different species of *A*-statements as follows:

$(A_p)^2$ Any statement which is not necessarily true (false) makes an *A*-determination of pastness if, and only if, it can be used to make a true statement only if used later than the event or state of affairs it reports.

$(A_{pr})^2$ Any statement which is not necessarily true (false) makes an *A*-determination of presentness if, and only if, it can be used to make a true statement only if used simultaneously with the event or state of affairs it reports.

$(A_f)^2$ Any statement which is not necessarily true (false) makes

[1] Usually when ordinary language philosophers or linguistic analysts speak of 'rules of use' they are referring to linguistic regularities that have an implicitly normative status in that the speakers of the language in question accept correction when they have deviated from these regularities and offer correction to others who deviate from them. These Truth-condition Rules of Use are rules in a stricter sense: because they are actually recorded in grammar books, they have a more explicitly normative status.

an *A*-determination of futurity if, and only if, it can be used to make a true statement only if used earlier than the event or state of affairs it reports.

By the event or state of affairs reported by a statement is meant the event or state of affairs referred to by the nominalization of this statement. For example, the event reported by '*S* is now ϕing' is the event referred to by '*S*'s ϕing', which is, of course, the event consisting of *S*'s ϕing. I shall assume that there can be a molecular state of affairs, so that there can be a state of affairs which is reported by a molecular statement. The molecular statement, '*S* is now ϕing or R is now ϕing', reports the state of affairs of *S*'s now ϕing or R's now ϕing, while the statement, 'R will ϕ if *S* is now ϕ', reports the state of affairs of R's going to ϕ if *S* is now ϕ. This way of talking is a bit awkward, but it is necessary if we are to apply these Truth-condition Rules for *A*-sentences to molecular sentences. I shall also count so-called 'negative events (states of affairs)' as events (states of affairs). The negative statement, '*S* is not now ϕing', reports the negative event of *S*'s not now ϕing; it qualifies as an *A*-statement making a determination of presentness since it is made through the use of a sentence that can be used to make a true statement only if used simultaneously with the event it reports, which in this case is a negative one.

We have now achieved the desideratum of finding a language-neutral criterion for an *A*-statement, as well as for different species of *A*-statements. As a result we are now in a position to determine, without begging any questions which English sentences, along with sentences in other languages, are used to make an *A*-statement. We shall first consider certain non-ordinary English sentences, and in particular those which occur in the *analysans* of the TRI analysis, since a great deal of philosophical importance turns on whether they are *A*-sentences. According to TRI the *A*-statement, '*S* is now (was, will be) ϕing', is analysed into, '*S*'s ϕing *is* simultaneous with (earlier than, later than) this token', in which 'this token' refers to the occurrence of the entire sentence token within which it occurs. According to criterion (*A*)[2] the statement in the *analysans* of this analysis is an *A*-statement, the reason being that it is made through the use of a sentence for which it is logically possible that two non-simultaneous uses of it make statements differing in truth-value, even if both statements refer to the same *things* and the same places. Notice that the requirement that the same *things* be referred to by each use of the sentence is fulfilled by any two uses of the sentence, '*S*'s ϕing *is* simultaneous with (later than, earlier than) this token', because 'this token' refers to an *event*—the token event consisting of the uttering or inscribing of the sentence token within which it occurs. Temporal relations can hold only between events: if the TRI *analysans* is to make sense the event consisting in *S*'s ϕing must be temporally related to the *occurrence* of a sentence token and not to a sentence token considered as an enduring object, such as ink marks on a piece of paper. Not only is the *analysans* of TRI an *A*-statement, it also is exactly the same species of *A*-statement as is the *analysandum*. This is due to the fact that the sentence used in making the statement in the *analysans* obeys the same Truth-condition Rule for Use of *A*-sentences as does the sentence which is contained in the *analysandum*: '*S* is now ϕing' and '*S*'s ϕing *is* simultaneous with this token' both satisfy criterion $(A_{pr})^2$. Thus the TRI analysis fails to analyse *A*-statements into *B*-statements.

We could imagine a language in which *A*-determinations are made by the colour in which a sentence is inscribed—red being past, white being present and blue being future—or the pitch at which a sentence is sung, this being practical only for a tribe of opera singers. In such cases our concept of a sentence type would have to include the colour or pitch of a sentence token, as well as the sign design and sound pattern. Criterion $(A)^2$ could also be extended to sign languages: an *A*-determinative gesture is one which involves a gesture type for which it is logically possible that two non-simultaneous performances of this gesture make statements differing in truth-value. There could be a language in which *A*-determinations are made by a combination of spoken utterances and gestures: if one utters a sentence while standing on his left foot it makes an *A*-determination of past, and so on. Herein a sentence type would have to include reference to both a gesture and a pattern of noises.

To return from imagined to actual languages, there are non-Indo-European languages that use grammatical devices for making *A*-determinations other than tensed verbs or copulae. For example, in Eskimo *A*-determinations are made by altering the end of substantives:

> In Eskimo *Ningla* 'cold, frost,' has a preterit *ninglithuk* and a future *ninglikak*, and from *puyok* 'smoke' is formed a preterit *puyuthluk* 'what has been smoke,' and a future *puyoqkak* 'what will become smoke.'[1]

Several American Indian languages use a similar grammatical device for making *A*-determinations.[2] In certain African languages like Mende and Hausa pronouns are inflected, instead of verbs or nouns, for the purpose of making *A*-determinations. 'The chief went' would be translated by words meaning 'Chief he-past go'.[3] The latter sentence translated into Mende or Hausa obeys the same Truth-condition Rule of Use as the English sentence 'The chief went'; each satisfies $(A_p)^2$ because it can be used to make a true statement only if used later than the chief's leaving.

Failure to distinguish between the particular grammatical technique employed by some language for making *A*-statements and the inter-linguistic rules of use which this technique exemplifies has led several philosophers and linguistic relativists to say silly things about the relativism of the way in which we conceive of time to the grammatical structure of the language we speak. They seem to believe that the only way in which an *A*-statement can be made is through the use of a tensed sentence. This is analogous to thinking that every king in chess must be made of ivory because in certain known chess sets the king is made of ivory. Donald Williams has claimed that Broad's conviction that the *A*-Series is objective 'results from the spell of Indo-European language, which bristles with time distinctions'.[4] Williams suggests as an antidote to the vicious spell cast over philosophers by the tenses of our language ridding our language of tenses by the TRI analysis! He fails to see that this is only substituting an alternative grammatical technique for making *A*-statements.

The particular manner in which a language makes *A*-statements may express or reinforce a particular metaphysical or theological view. For instance, there could be a language in which the only

[1] Otto Jespersen, *The Philosophy of Grammar*, p. 283.

[2] F. Boas, *Handbook of American Indian Language*, p. 105 ff.

[3] G. W. Turner, 'Time and Place Logic', *Philosophy*, XXXVI (1961). Unfortunately, Turner draws the wrong conclusion from this—that persons who speak a language such as Mende or Hausa have a different conception of time from those who speak an Indo-European language.

[4] Williams, 'The Sea Fight Tomorrow', in *Structure, Method and Meaning*, edited by Henle, Kallen and Langer (New York, 1951), p. 299.

way to say that something happened (is happening, will happen) is by affixing to statements the tensed theological operator 'it has been revealed by God that (is now being revealed by God that, will be revealed by God that)'. Benjamin Lee Whorf, I believe, fails to distinguish between the question whether a given language has devices for making *A*-statements and the question concerning what metaphysical commitments are packed into the devices by which a language makes *A*-statements. He claims that 'the Hopi language contains no reference to "time", either explicit or implicit', and in the very next sentence goes on to say that 'the Hopi language is capable of accounting for and describing correctly, in a pragmatic or operational sense, all observable phenomena of the universe'.[1] This appears to be inconsistent. Among the 'observable phenomena' are *temporal* facts, such as that *S* is ϕing at present and that event *M* immediately preceeds *N*. Since Hopi can describe all observable phenomena, it must be able to describe these temporal facts. But then it is false that this language 'contains no reference to time, either explicit or implicit'.

It turns out, on Whorf's own account, that Hopi does after all have devices for making temporal determinations, including *A*-statements. Hopi contains verb forms, which translated into English mean something like the 'Manifested and Manifesting'.

> The objective or manifested comprises all that is or has been accessible to the sense, the historical universe, in fact, with no attempt to distinguish between present and past, but excluding everything that we call future.[2]

The manifesting refers to *present* tendencies of a dynamic sort. Thus Hopi does have devices for making *A*-statements. This is quite consistent with Whorf's further claim that the Hopi concepts of the Manifested and the Unmanifested express an animistic metaphysics. This is interesting but totally irrelevant to his earlier claim that 'the Hopi language contains no reference to time'.

[1] Whorf, 'An American Indian Model of the Universe', in Whorf's *Language, Thought and Reality*, edited by J. B. Carroll (Cambridge, Mass., 1956), p. 58. This article is reprinted in *The Philosophy of Time*, *op. cit.*

[2] *Ibid.*, p. 59. A Hopi sentence containing a verb form expressing what is manifested satisfies criterion $(A)^2$, since it is logically possible for two non-simultaneous uses of this sentence to make statements differing in truth-value. While this sentence is an *A*-sentence, it does not satisfy either $(A_p)^2$ or $(A_{pr})^2$ or $(A_f)^2$. There also are English sentences that satisfy $(A)^2$ but fail to satisfy any one of the latter criteria, for example, '*S* has always been ϕing up to the present'.

At best, he shows that the way in which Hopi makes *A*-statements carries with it a certain metaphysical view. Of course, there might be a tribe hiding out somewhere whose 'language' really contained no devices for making *A*-statements. Should this group of primitive chaps be discovered some day, they, fortunately, would not constitute a counter-example to anything said above.

Before going on to consider the criterion for a *B*-statement, there is a curious feature of our criterion for an *A*-statement, as well as those for the different species of *A*-statements, which warrants our attention. This curious feature concerns the fact that criterion $(A)^2$, along with $(A_p)^2$, $(A_{pr})^2$ and $(A_f)^2$, make use of *B*-expressions. $(A)^2$ speaks of two *non-simultaneous* uses (i.e. uses at an *earlier* and *later* time) of a sentence, while $(A_p)^2$, $(A_{pr})^2$ and $(A_f)^2$ specify that the use of the sentence must be respectively *later than*, *simultaneous with* and *earlier than* the event reported if a true statement is to be made. Someone might seize upon this to support the *B*-Theory's claim that *A*-determinations are analysable in terms of *B*-relations. The claim here would be that when *A*-determinations are *analysed*, *B*-expressions are used in the *analysans*. This way of supporting the *B*-Theory contention is inadequate for two reasons.

It is inadequate, first, because it misconstrues the *B*-Theory contention that, to speak in the linguistic idiom, *A*-expressions are analysable solely in terms of *B*-expressions, which is a claim about the *object* language. According to this contention, an *A*-statement can be analysed in terms of an equivalent *B*-statement. The Linguistic Reduction of the *B*-Theory attempted to show that a statement containing an *A*-expression could be analysed into an equivalent statement containing *B*-expressions exclusively. Criteria $(A)^2$ and those for the various species of *A*-statements are based on the rules of use of *A*-sentences, which rules are concerned with the conditions under which these sentences can be used to make true (false) statements. The Truth-condition Rules for the Use of *A*-sentences, like all rules of use, are formulated in the *metalanguage*. No *B*-Theorist would be so foolish as to claim that the object language statement, '*S* is now ϕ', is equivalent to the metalinguistic statement, 'The sentence, "*S* is now ϕ", can be used to make a true statement only if used simultaneously with *S*'s being ϕ.' The former statement refers to *S* and informs us about what attribute it has now, whereas the latter tells us nothing about

S but rather something about the conditions under which a certain sentence can be used to make a true statement. To think these statements are equivalent is analogous to thinking that there is no difference between describing the rules for the use of the king in chess and making a move with the king in some chess game.

This claim is inadequate, in the second place, because the criterion for an *A*-statement, as well as the criteria for different species of *A*-statements, need not utilize any *B*-expression. These metalinguistic criteria can be rephrased so that they use *A*-expressions exclusively. For example, criterion $(A)^2$ for an *A*-statement can be reformulated as:

$(A)^3$ Any statement which is not necessarily true (false) is an *A*-statement if, and only if, it is made through the use of a sentence for which it is possible that it is now used to make a true (false) statement and some past or future use of it makes a false (true) statement, even if both statements refer to the same things and the same places.

Criteria $(A_p)^2$, $(A_{pr})^2$ and $(A_f)^2$ can be similarly rephrased respectively as:

$(A_p)^3$ Any statement which is not necessarily true (false) makes an *A*-determination of pastness if, and only if, it is made through the use of a sentence which can now be used to make a true statement only if the event it reports is now past.

$(A_{pr})^3$ Any statement which is not necessarily true (false) makes an *A*-determination of presentness if, and only if, it is made through the use of a sentence which can now be used to make a true statement only if the event it reports is now present.

$(A_f)^3$ Any statement which is not necessarily true (false) makes an *A*-determination of futurity if, and only if, it is made through the use of a sentence which can now be used to make a true statement only if the event it reports is now future.

A disjunction of these three criteria for different species of *A*-statements can be used for giving a new criterion for an *A*-statement:

$(A)^4$ Any statement which is not necessarily true (false) is an *A*-statement if, and only if, it is made through the use of a sentence which satisfies either $(A_p)^3$ or $(A_{pr})^3$ or $(A_f)^3$ and does not satisfy more than two of them.

The 'either-or' in this criterion is inclusive. A temporally ubiquitous statement fails to satisfy $(A)^4$ because it is made through the use of a sentence which satisfies all three of the mentioned criteria —$(A_p)^3$, $(A_{pr})^3$ and $(A_f)^3$. A statement of the form, '*S* has always been ϕing up to the present', qualifies as an *A*-statement because the sentence used in making this statement satisfies only two of these criteria—$(A_p)^3$ and $(A_{pr})^3$.[1] One advantage of $(A)^4$ over $(A)^3$ is that it is not necessary to restrict $(A)^4$ so as to rule out sentences that are non-freely repeatable solely in virtue of containing a non-temporal indexical word. In the remainder of this book I shall use $(A)^3$ because it is less unwieldy. But it is still comforting to have $(A)^4$ in reserve just in case $(A)^3$ does not successfully rule out sentences containing only non-temporal indexical words.

Some are bound to object that criteria $(A)^3$, as well as $(A_p)^3$, $(A_{pr})^3$ and $(A_f)^3$, are viciously circular, since in giving, for instance, the criterion for an *A*-statement the *A*-expressions 'now', 'past', and 'future' are used. All this shows, on the contrary, is that in order to be able to understand the criterion for an *A*-statement one must be able to understand some *A*-statements. Similarly, in order to understand the criterion for the truth of a conjunction to the effect that a conjunction is true if, and only if, each one of its conjuncts is true one must be able to understand some conjunctions.[2]

We now turn to the task of finding a language-independent criterion for a *B*-statement. Since the aim of most *B*-Theorists is the creation of an ideal language, a *B*-statement ought to be freely repeatable. This suggests that a *B*-statement is made through the use of a sentence which makes a statement having the same truth-value whenever it is used. This gives only a necessary condition for being a *B*-statement, for there are freely repeatable sentences, such as 'Two and two *is* four', which obviously are not used to

[1] A Hopi statement about some state of affairs being manifested is ambiguous as to whether this state of affairs is past or present. Nevertheless, this statement satisfies $(A)^4$.

[2] I am indebted to Nuel Belnap for suggesting this way of countering the charge of vicious circularity.

make a *B*-statement. We might try to make this criterion based on free repeatability sufficient as well as necessary by adding the further requirement that a *B*-statement must also make a *temporal* determination. A temporal determination, supposedly, is a genus whose two species are *A*-determinations and *B*-relations. But it is not clear that these are the only two species of temporal determinations. For example, do not '*S is* ϕ at some time' and '*S* is always ϕ' also make 'temporal determinations'? They are neither *A*- nor *B*-statements, even though they are made through the use of a freely repeatable sentence.

The following criterion for a *B*-statement overcomes this ambiguity of 'temporal determination':

(*B*) Any statement is a *B*-statement if, and only if, it describes a temporal relation between events and is made through the use of a sentence for which it is the case that if it can now be used to make a true (false) statement then any past or future use of this sentence also makes a true (false) statement.

Once again it might be charged that our criterion is circular, because it uses the expression 'temporal relation' in giving the criterion for a *B*-statement.[1] Criterion (*B*) is language-independent because it makes no mention of any items of vocabulary or grammar that must be contained in a sentence that is used to make a *B*-statement. It can now be determined whether a given sentence, be it in English or some other language, is used to make a *B*-statement.

The following sort of objection might be raised against our language-independent criteria for *A*- and *B*-statements. Both of these criteria are based upon the Truth-condition Rules of Use for the sentences employed in making *A*- and *B*-statements. According to these criteria, *A*- and *B*-statements are made through the use of sentences that obey different Truth-condition Rules of Use. But if two sentences obey different Truth-condition Rules of Use they cannot be used to make the same or equivalent statements. Therefore, our criteria for *A*- and *B*-statements begs the question against the *B*-Theory's claim that an *A*-statement can be translated without loss of meaning into a *B*-statement.

This objection is unfounded because that two sentences obey

[1] *Simultaneous with*, as stipulated in footnote 1 on page 17, is to be counted as a *B*-relation, and therefore it also qualifies as a temporal relation.

different Truth-condition Rules of Use does not entail that they cannot be used to make equivalent statements. The sentences 'I am six feet tall' and 'You are six feet tall', although subject to different Truth-condition Rules of Use, the former making a true statement only if used by a person six feet tall and the latter only if the person addressed by the speaker is six feet tall, can be used to make the same or equivalent statements. This would happen if these two sentences were used at the same time and referred to the same person. Therefore, it is possible for sentences subject to different Truth-condition Rules of Use to be used to make the same statement. It is also possible that sentences obeying identical Truth-condition Rules of Use can never be used to make the same or equivalent statements. An example of such a pair of sentences are those employed in the *analysandum* and the *analysans* of the TRI analysis—'*S* is now ϕ' and '*S*'s being ϕ *is* simultaneous with this token'. As seen above, they satisfy the same Truth-condition Rules of Use, since each can be used to make a true statement only if used simultaneously with *S*'s being ϕ. Nevertheless, the statements made by the simultaneous uses of these sentences are never logically equivalent. The reason is that the statement that *S* is now ϕ, unlike the statement that *S*'s being ϕ *is* simultaneous with this token, does not entail that there is a token.[1]

[1] Statements having different entailment relations cannot be logically equivalent. This topic will be explored in great detail in Chapter XI.

IV

THE INELIMINABILITY OF *A*-DETERMINATIONS

THE criteria for *A*- and *B*-statements developed in the preceding chapter can now be put to work in helping to mediate the dispute between the contrary claims of the *A*- and *B*-Theory in their first tenet. Tenet (1) of the *B*-Theory, which holds that the *A*-Series is reducible to the *B*-Series since *A*-determinations are analysable in terms of *B*-relations, will come under fire in this chapter. Arguments will be advanced to show that *A*-determinations are not reducible to *B*-relations. Even if successful, these arguments will not establish the *contrary* tenet (1) of the *A*-Theory—that the *B*-Series is reducible to the *A*-Series since *B*-relations are analysable in terms of *A*-determinations. It might turn out that neither the *A*-Series is reducible to the *B*-Series nor the *B*-Series to the *A*-Series. The only tenable theory of time would be the 'Neither-Way-Will-Work Theory', which holds that both *A*- and *B*-sentences must be used to describe all the temporal facts there are.[1]

Let us quickly review the various techniques employed by the *B*-Theory for analysing *A*-determinations in terms of *B*-relations. All of these techniques involved analysing an *A*-statement into a *B*-statement. The Linguistic Reduction of the *B*-Theory contained two steps: the first step, TRI, making explicit in the *analysans* the token-reflexivity of *A*-expressions; and the second step, TRII,

[1] The Neither-Way-Will-Work Theory is defended by Louis Mink in 'Time, McTaggart and Pickwickian Language', *The Philosophical Quarterly*, 10 (1960).

eliminating in the *analysans* all token-reflexive words. The TRI analysis of '*S* is now ϕ' into '*S*'s being ϕ *is* simultaneous with this token' fails to analyse an *A*-statement into a *B*-statement since its *analysans* is an *A*-statement according to any one of the three criteria for an *A*-statement—$(A)^2$, $(A)^3$ or $(A)^4$. The TRII analysis of '*S* is now ϕ' into '*S*'s being ϕ *is* simultaneous with *theta*', in which '*theta*' is a metalinguistic proper name for the occurrence of the tensed sentence token in the *analysandum*, does succeed in having a *B*-statement in its *analysans* according to criterion (*B*) for a *B*-statement: it is logically impossible that the sentence employed in its *analysans* now be used to make a true (false) statement and some past or future use of it make a false (true) statement. A variation on the TRII analysis cashes in the metalinguistic proper name '*theta*' in its *analysans* for a definite description of the token event in the *analysandum*—'*S*'s being ϕ *is* simultaneous with the occurrence of a token of "*S* is now ϕ".' This latter statement also is a *B*-statement according to (*B*): it mentions but does not use the tensed sentence '*S* is now ϕ'. By placing this sentence within inverted commas, it helps to identify the token event in the *analysandum* by giving an icon of the sentence token which occurs in this token event.

The vital question is whether the *analysans* of the TRII analysis and its variation are an adequate analysis of the *analysandum*. What are the conditions that must be fulfilled by an adequate analysis of one statement in terms of another? It is not sufficient that the *analysandum* and *analysans* be materially equivalent. If in fact the one and only token of '*S* is now ϕ' occurs simultaneously with *S*'s being ϕ, it establishes the material equivalence of the *analysans* and *analysandum* of the TRII analysis and its variation, but this is far too weak to ensure that the *analysans* gives an adequate analysis of the *analysandum*. That grass is green is materially equivalent to that snow is white, both being true, but no one would want to say that either one is an analysis of the other. Some stronger condition than material equivalence must be laid down. Such a stronger condition would require that the *analysans* entail the *analysandum*. An even stronger condition would require that they be logically equivalent.[1]

TRII and its variation fail to satisfy either one of these further

[1] There are stronger conditions for an adequate analysis, but they will not be pursued here.

conditions for an adequate analysis. The *B*-statements in the *analysans* of these two analyses do not entail the *A*-statement in the *analysandum*: that *S*'s being ϕ *is* simultaneous with *theta* (the occurrence of a token of '*S* is now ϕ') does not entail that *S* is now ϕ. These *B*-statements describe a *B*-relation between *S*'s being ϕ and a certain token event without entailing that either one of these events is now present (past, future). That they do not convey or entail information about the *A*-determination of an event can be seen by the fact that whenever someone uses the sentence, '*S*'s being ϕ *is* simultaneous with *theta* (the occurrence of a token of "*S* is now ϕ"),' he has not forestalled the question whether *S*'s being ϕ (or the occurrence of *theta*) is now present (past, future). The same considerations hold for the *B*-Theory's attempt to analyse an *A*-statement into a *B*-statement through the tenseless ascription of dates, e.g. '*S is* ϕ at t_7.' The latter statement describes a *B*-relation between *S*'s being ϕ and the event seven time-units earlier than it which serves as the origin of the calendar, but does not entail that *S*'s being ϕ or t_7 is now present. Tensed or *A*-questions of the form 'Is *S* now ϕ?' and 'What is the date (time) now?' can be answered only by contingent *A*-statements. No amount of information about the *B*-relations between *S*'s being ϕ and other events will answer this tensed question.

Since the *analysans* of the above analyses do not entail the *A*-statement in the *analysandum*, it *ipso facto* establishes that these analyses also fail to satisfy the stronger condition for an adequate analysis—that the *analysans* and *analysandum* be logically equivalent. That their *analysans* and *analysandum* are not logically equivalent is due to the fact that the *analysans* could be true (false) and the *analysandum* false (true). It could be the case that it is true (false) that *S* is now ϕ and be false (true) that *S*'s being ϕ *is* simultaneous with *theta* (t_7).

That a statement containing an *A*-expression cannot be translated into a *B*-statement without loss of factual or informative content suggests that *A*-expressions are used to convey factual information, namely information about the *A*-determination of an event or state of affairs. Information about an event's *A*-determination seems to be part of the content of what is asserted or stated by someone who uses an *A*-sentence, since one of the truth-conditions of his statement is that the event reported by his *A*-statement has the specific *A*-determination it is asserted to have. For example, if I make the *A*-statement that Jones is dead, you could say that my statement is false because Jones is not yet dead: his death, in other words, has the *A*-determination of futurity rather than pastness, as I claimed.

Before exploring the question of the logical status of the kind of information conveyed by *A*-expressions, it is a good idea to drive home the fact that *A*-expressions are ineliminable because they do convey some kind of factual information. This can best be done by giving examples of the informative role that *A*-sentences play in our daily lifes in which we carry on our commerce and traffic with the world and our fellow man. Let us imagine some situation in which language is used as an instrument to enable two or more men to communicate with each other so that they can co-ordinate their actions in the achievement of some shared goal. Joe is a scout for a machine-gun company. He is strategically stationed so that he can survey the battlefield, and when the enemy approaches within 100 yards of their position he must inform the company of this fact. The company commander will then collate this piece of information with other information, such as whether enemy fighter planes are then overhead and could spot their position if they fired, and then decide whether or not to give the company the order to open fire.

The crucial question concerns whether Joe can alert the company commander of the fact that the enemy is now within a 100-yard range through the use of a *B*-sentence. There is no question of Joe being able to inform the commander of this fact by the use of an *A*-sentence, such as 'The enemy is now within 100 yards', 'The enemy's being within 100 yards *is* simultaneous with this token,' which is the *analysans* of TRI. Joe might have used a conventionally agreed upon *A*-gesture, such as holding his rifle above his head, which gesture would play the same role as one of the above *A*-sentences. This gesture would be temporally restricted, such that if Joe held his rifle above his head when the enemy was not within a 100-yard range he could be charged with saying something false.

Could Joe inform the commander of the present advance of the enemy through the use of a *B*-sentence? It seems obvious that by using a sentence which satisfies criterion (*B*) he could not impart this information to the commander. Suppose he spots the enemy within the 100-yard range and then whispers to himself the sentence, 'The enemy is now within 100 yards.' Let us give the metalinguistic proper name '*theta*' to this token event. Immediately upon whispering this sentence token to himself, Joe says out loud the *B*-sentence, 'The enemy's being within 100 yards *is* simultaneous with *theta* (the utterance by Joe of "The enemy is now within 100 yards"),' which is the *analysans* of TRII (TRII's variation). This *B*-statement does not convey or entail any information about the *A*-determination of the state of affairs of the enemy's being within 100 yards: it merely describes a *B*-relation of simultaneity between this state of affairs and the occurrence of a particular sentence token.

Against this Bernard Mayo has argued that information about an event's *A*-determination could be conveyed through the use of a freely repeatable sentence. He attempts to support this by translating the *A*-statement, 'Atlas is firing now,' into the *B*-statement, 'Atlas' firing *is* simultaneous with the utterance of "Atlas' firing is . . .".'[1] If 'the utterance of "Atlas' firing is . . ."' is a description of a linguistic event consisting in someone inscribing or uttering a token of 'Atlas' firing is . . .' then Mayo's *analysans* is a *B*-statement according to (*B*) but it fails to convey information about the firing of the Atlas occurring now. If, on the other hand, 'the utterance of "Atlas' firing is . . ."' is meant to be token-reflexive, then the *analysans* does convey this information, but it no longer satisfies (*B*); instead it satisfies $(A)^2$ and more specifically $(A_{pr})^2$. Another of Mayo's counter-examples, which also fails, is the use of 'Hurrah! Atlas' firing *is* simultaneous with the utterance of "Hurrah".'[1] This entire unit of discourse—exclamation with following sentence—is not a *B*-sentence, since it can be used to make a true statement only if used simultaneously with the firing of the Atlas. Neither the use of the exclamation by itself nor the use of the sentence following it by itself conveys information about the *A*-determination of the firing of the Atlas.

Possibly Joe could inform the commander of the present advance of the enemy through the use of a *B*-sentence tenselessly ascribing a date to this event. For example, after spying the enemy within 100 yards and checking his calendar-watch, he could say, 'The enemy *advances* within 100 yards on December 12, 1966, at

[1] B. Mayo, 'Infinitive Verbs and Tensed Statements', *The Philosophical Quarterly*, 13 (1963), p. 296.

[1] *Ibid.*

4:01 p.m. E.S.T.' The statement made through the use of this sentence fails to convey information about the *A*-determination of the event reported, since the company commander, upon hearing Joe's utterance, must check his own calendar-watch so as to find out that the date, and therefore the dated event, is now happening. Before elaborating on this, a misguided attempt to show that Joe's statement fails to convey information about the present advance of the enemy will be considered.

The misguided objection to the claim that Joe's *B*-statement conveys the same information as the *A*-statement, 'The enemy is now within 100 yards', is that there are practical difficulties attendant upon the former which do not arise for the latter, having to do with the often disastrous time lapse involved in Joe and the company commander having to check their calendar-watches. In many situations there is not sufficient time for the communicants to consult their calendar-watches. For example, I see the roof caving in on you and wish to alert you to this fact; but I first must check my calendar-watch and then say something like, 'The roof *caves* in on you at t_7,' and upon hearing this you, by now buried by plaster and beams, must check your calendar-watch. Almost as disastrous is the situation in which a woman says to her lover at the appropriate time, 'Kiss me now!'; even calendar-watches with luminous dials will not be much consolation.

A *B*-Theorist could make the justified reply to this that he is not creating a language for lovers and warriors. In any philosophically illuminating analysis the *analysans* of the analysis will be longer, and more cumbersome, than the *analysandum*; and as a result there will be practical difficulties in using the *analysans* rather than the *analysandum* in daily discourse. But this does not show that the *analysans* does not contain the same factual content as the *analysandum*. Wittgenstein said:

> Language is an instrument. Its concepts are instruments. Now perhaps one thinks that it can make no *great* difference *which* concepts we employ. As, after all, it is possible to do physics in feet and inches as well as in metres and centimetres; the difference is merely one of convenience. But even this is not true if, for instance, calculations in some system of measurement demand more time and trouble than it is possible for us to give them.[1]

[1] L. Wittgenstein, *Philosophical Investigations* (Oxford, 1953), # 569.

This is true enough, but it would not show that a philosophical analysis of the metric concepts of physics into more cumbersome and less familiar concepts, which therefore would lack the utility of the former, would be incorrect. All that the analyst contends is that everything we say with the *analysandum* could be said, although not with the same fluidity and convenience, by the analysans.[2]

The reason why Joe's *B*-statement does not convey information about an event's *A*-determination has nothing to do with any practical inconvenience that it causes. Rather it is because the statement Joe makes, namely that the enemy *advances* within 100 yards on December 12, 1966, at 4:01 p.m. E.S.T., does not convey information about the *A*-determination of the reported event, nor does it entail a statement that conveys such information. If it did, it would not be necessary for the company commander to do something else after he hears and understands Joe's utterance—checking his calendar-watch to find out if the dated event has the *A*-determination of presentness. The general principle here is that if a statement conveys an item of information *alpha*, which then makes it possible for the hearer to discover on his own through non-logical means an additional item of information *beta*, then this statement has not conveyed item of information *beta*: *beta* is not part of its factual or informative content or, in other words, not part of the statement made. If I inform you of the name of a girl you met at a party last night, and this item of information enables you to discover *on your own* through non-logical means her phone number and address by consulting a telephone directory, we would not say that I communicated to you information concerning her phone number and address. And, if you should call her and make a date, we certainly would not say that I fixed you up with her. And if things should work out between you and the girl and eventually you marry, we would not say that I arranged your marriage. Therefore, it is incorrect to say that a speaker who

[2] There is another reason why I do not want to claim that a *B*-statement cannot be an adequate translation of an *A*-statement because there are practical difficulties connected with the former. In Chapter VI I shall attempt to translate a *B*-statement into a disjunction of *A*-statements; but one could argue that such a reduction fails because there are some contexts, such as science, in which it is *practically* advantageous to use only *B*-sentences. The success of my reduction of *B*-statements to *A*-statements rests on the fact that everything we say through the use of *B*-sentences could be said through the use of *A*-sentences. Practicality has nothing to do with it.

makes a *B*-statement tenselessly ascribing a date to an event communicates information concerning the *A*-determination of this event, even though this information makes it possible for the hearer to discover by non-logical means whether this dated event has a certain *A*-determination.

Irving Thalberg, who has argued that Joe's *B*-statement could convey information about an event's *A*-determination, has argued as follows:

> If someone objects that I beg the question by inserting a tensed statement into the listeners' mouth or mind (when he checks his calendar-watch), my answer would be: there is no difficulty in supposing the machine-gunners react appropriately to Joe's warning but their response is unaccompanied by overt or sub-vocal translations. The gunners may have been trained to do so; just as a tailor's apprentice may be taught to fetch Harris tweeds or herringbones, in response to the tailor's commands, without first looking at swatches of cloth or his memory-images of Harris or herringbone tweed.[1]

First, whether or not the hearer utters an *A*-statement, either to himself or out loud, when he checks his calendar-watch, does not alter the fact that he discovers some new item of information not given him by the speaker nor entailed by what the speaker has said.[2] Our criterion for saying this might be merely that before he checked his calendar-watch he could not answer correctly the question of whether the dated-event is present, whereas afterwards he could. Second, while we know what it is like for a chart correlating words with material swatches to be an instrument involved in teaching someone how to play the 'language-game' of fetching different material when ordered to do so and later to drop out (no longer be consulted by the apprentice) in the playing of this game when memory or habit takes over, we have no idea what it would be like for the calendar-watch to be an instrument *only* in teaching someone to play the language-game of discovering whether some dated-event is present and later to drop out in the

[1] Thalberg, 'Tenses and "Now"', *The Philosophical Quarterly*, 13 (1963), p. 310.

[2] While the hearer need not utter an *A*-statement when he checks the calendar-watch, one could argue that nevertheless an *A*-statement is made, only it is made by the calendar-watch. When we ask someone to check their watch and tell us the time we often say, 'What time does *it* (the watch) *say*.' The clock-readings are *A*-sentences. Just as we have clocks which tell the time by flashing the time in lights we could have a clock which *told* the time by uttering *A*-sentences, such as 'It is now t_7'.

playing of this game as memory or habit takes over. Thalberg's comparison of these two different language-games falsely suggests that someone might be trained so as to know through memory or habit what the date is!

One might argue that it is only a contingent fact that men do not possess a faculty for knowing through immediate intuition what the date is, and if they were to possess such a faculty a *B*-statement tenselessly ascribing a date to an event could have the same informative content as an *A*-statement. There could be a race of chrono-sense-datumists who did not have to inspect external calendar-watches to know what the date and time is: each member of this race would have a built in calendar-watch in his 'mind's eye' of which he is always immediately aware through introspection.[1] For this race of people *A*-expressions would add nothing to the informative content of a *B*-sentence tenselessly ascribing a date to an event.[2] But this fanciful objection proves too much, for by the same sort of counter-factual suppositions it could be shown that no word has an informative function. For instance, why not assume that this race of people has an ability to intuit with their mind's eye the colour of any object referred to; and thus they would have no more need for colour words in their language than they do for *A*-expressions.[3] But certainly this does not show that colour words and *A*-expressions do not have a factual or informative role in *our* language. It shows only that these amazingly endowed persons would have no need to employ such words in their language. No doubt our concept of factual or informative meaning is dependent upon the way in which we experience the world—that certain things can be discovered only through sense experience—and if our epistemic relation to the world were to be radically different, in the manner just described,

[1] I owe this example to the wild imagination of John King-Farlow.

[2] There is only one drawback to these 'inner' clocks. If some member of this race should be hit on the head and then find that his 'inner' time readings no longer agree with those of the majority it would require a major operation to synchronize his clock with the others.

[3] This is not completely accurate. These people would have an initial need for *A*-expressions, since they would have to find out that they agreed upon what date and time it is now. The use they make of *B*-sentences ascribing a date to an event presupposes this agreement in judgment as to the present date. Since they cannot inspect each other's inner clock they can determine that they agree in their judgments only by uttering *A*-sentences and discovering that the statements made are in agreement.

we very likely would have a different concept of factual or informative meaning. But the question, 'Do *A*-expressions have a factual or informative role?', is raised within our language and so has reference to our concept of factual meaning—not the one we might have if we were a radically different sort of being.

Nelson Goodman has claimed that an *A*-statement can be translated without loss of meaning into a *B*-statement.

> Since—the argument runs—we recognize the tenseless sentence as a translation of the tensed one only in the light of outside knowledge, we have no genuine translation at all. But this seems to me no more cogent than would be the parallel argument that 'L'Angleterre' is not a genuine translation of 'England' because we recognize it as a translation only if we know that L'Angleterre is England.[1]

'England' can be a genuine translation of 'L'Angleterre' only because these two words obey the same rules of use: they play analogous roles in their respective languages. Since an *A*- and a *B*-sentence obey different rules of use one could not be a genuine translation of the other. 'I am ill' is a genuine translation of 'Ich bin krank', but 'I *am* ill in 1966' is not. And, for the same reason, 'I *am* ill in 1966' is not a genuine translation of 'I am ill'. 'England' and 'L'Angleterre' have the same meaning (i.e. the same sense) and therefore one is substitutable for the other in any extensional context without changing the truth-value of the statement made; however, an *A*- and a *B*-sentence cannot have the same meaning since the substitution of one of them for the other in some extensional context could change the truth-value of the statement that is made.

So far the thesis that an *A*-statement cannot be rendered by a *B*-statement without loss of factual meaning has been defended against numerous counter-examples. But this does not give us the right to conclude that *only* an *A*-statement can convey or entail information about an event's *A*-determination. Possibly such information could be conveyed or entailed by a statement which is neither an *A*- nor a *B*-statement. Jay Rosenberg has attempted to give an example of such a statement:

> The charging of the Light Brigade *is* simultaneous with the one and only one utterance the first ten words of which are 'The charg-

[1] Goodman, *The Structure of Appearance, op. cit.*, p. 297.

ing of the Light Brigade *is* simultaneous with' and the last five words of which are '4 : 17 P.M., October 3, 1835,' and the charging of the Light Brigade *is* simultaneous with 4 : 17 P.M., October 3, 1835.[1]

This is definitely not a *B*-statement, since it is made through the use of a sentence for which it is not the case that if it can now be used to make a true statement then any past or future use of it will also make a true statement. Because the Rosenberg-sentence contains a *definite* description of itself, if there is more than one use of it a false statement will be made on each occasion. Neither is it an *A*-sentence according to criterion $(A)^2$ or its tensed paraphrase $(A)^3$, since it is not logically possible that it be used at two non-simultaneous times to make statements differing in truth-value or that it now be used to make a true statement and some past or future use of it make a false statement. This sentence does satisfy criterion $(A)^4$ based on the disjunction of the criteria for different species of *A*-statements, for it satisfies $(A_{pr})^3$, in that it can now be used to make a true statement only if the event it reports is now present. For the sake of argument, let us not take the Rosenberg-sentence to be an *A*-sentence.

This seeming counter-example to the thesis that only an *A*-statement can convey or entail information about an event's *A*-determination rests on a failure to distinguish between the statement made through the use of a sentence—what is said—and what can be discovered or inferred by the hearer from an inspection of the utterance token and other contextual circumstances of the speech act. If I use the sentence, 'It is now raining', I make the statement *that it is now raining*. The hearer, by inspection of my utterance token, can infer that my utterance token is in English and that I am able to speak English; however, I have not stated that my utterance token is in English nor does my statement—what I have said—entail that I can speak English. Someone who uses the Rosenberg-sentence does not state, nor does what he state entail, that his utterance token fits a certain definite description; although by uttering this sentence he makes it possible for the hearer to determine by inspection of his utterance token that it fits the definite description in question.[2] The speaker does not

[1] J. F. Rosenberg, 'A Note on Tensed Discourse and the Eliminability of Tenses', *The Philosophical Quarterly*, 16 (1966), p. 149.

[2] Thomas Schwartz, in an unpublished paper, 'Tense and Nontense', has created a language in which sentences that are neither *A*- nor *B*-sentences supposedly can be

state or *assert* that his utterance token fits the definite description in question but rather *shows* that it does, just as someone who uses the sentence 'It is now raining' does not *state* (nor does what he state's entail) that his utterance token is in English but *shows* that it is. If I should hold before your face a large calendar-watch when I use the sentence 'It *rains* at t_7' I would not have stated, nor would what I have stated entail, that it is now t_7, but rather I would have shown that this is so.

To summarize: the ineliminability of *A*-determinations has been defended against the attempts of *B*-Theorists to analyse *A*-expressions into *B*-expressions. By dwelling exclusively on whether *A*-expressions are *verbally* definable or analysable in terms of *B*-expressions, some other sense in which *A*-determinations are reducible to *B*-relations may have been overlooked. One *B*-Theorist, A. J. Ayer, has contended that *A*-determinations are *ostensively* definable solely in terms of *B*-relations, which can in turn be ostensively (and verbally) defined without reference to *A*-determinations.

> It can, I think, plausibly be maintained that the relation of temporal precedence is 'given' to us in experience. As a matter of empirical fact, one can see or hear *A*-following-*B*, in the same immediate fashion as one can see *A*-to the left of-*B*. And this relation of temporal precedence, coupled with the notion of the present, which may be defined ostensively, is all that is required to yield the concepts both of the past and of the future. Defining the present as the class of events which are contemporaneous with *this*, where *this* is any event that one chooses to indicate at the given moment, one can define the past as the class of events which are earlier than the present.[1]

[1] Ayer, *The Problem of Knowledge* (London, 1956), p. 152.

used to convey information about an event's *A*-determination. Each speaker of this language is assigned a unique number which he must affix to every one of his utterance tokens as a subscript and he also must place as a superscript to every one of his utterance tokens a number indicating what number utterance token of his it is. If I am given the number 7, then my first utterance token would look like, 'S is $\phi_7{}^1$', and my second utterance token would look like, 'S was $\phi_7{}^2$', and so on. He claims that the sentence, 'A Nubian's nod *is* simultaneous with the utterance token whose subscript is 7 and whose superscript is $9_7{}^9$', can be used to convey or entail that a Nubian is nodding at present. This counter-example suffers from the same weakness as the Rosenberg one: someone who uses the Schwartz-sentence does not state, nor does what he state's entail, that his utterance token has the subscript 7 and the superscript 9, although by uttering this sentence he does make it possible for the hearer to learn by inspection of his token that it has these scripts.

Ayer seems to suggest that *earlier than* (precedence) can be ostensively defined. Let us grant that it can, though many may be sceptical that anyone could be that fast of hand and tongue. One problem of Ayer's ostensive definition of the present is that it makes use of the *B*-relation *simultaneous with* (contemporaneous with), but Ayer nowhere ostensively or verbally defines this term. This flaw can easily be remedied: after ostensively defining *earlier than* we can verbally define 'simultaneous' in terms of earlier than by saying that two events are simultaneous if, and only if, neither one of these events is earlier than the other.[1]

There obviously is something radically wrong with Ayer's ostensive definitions of *A*-determinations, but to locate exactly where it is requires that a few preliminary things be said about ostensive definitions. An ostensive definition, as this term of jargon has been used in philosophy, is an attempt to answer a question of the form, 'What is *G*?', in which '*G*' is a general word for an attribute or kind, by pointing to something and saying, 'This is *G*.' Because an ostensive definition is of a sortal- or attribute-term it is unrestricted in its scope, applying to all entities, at all places and all times. An ostensive definition, such as 'This colour is red', is more perspicuously formulated as, 'Any thing having the same colour as this is red', since this brings out its unrestricted scope. Let us give the proper name '*A*' to the thing denoted by the use of 'this' in the giving of the preceding ostensive definition. Because the ostensive definition is temporally (as well as spatially) unrestricted, it entails that if at any time a thing has the same colour as *A* then at that time it is red.[2] There also is a negative way of formulating the temporal unrestrictedness of an ostensive definition: if at any time a thing does *not* have the same colour as *A* then at that time it is *not* red. We shall work with the negative formulation of the principle of temporal unrestrictedness and shall now show that when it is applied to Ayer's ostensive definitions of *A*-determinations it leads to a contradiction.

Ayer ostensively defines the present 'as the class of events which are contemporaneous with *this*, where *this* is any event that one

[1] Another problem with ostensively defining the present, which was indicated on page 5, is that necessarily everything pointed at in giving an ostensive definition is present, and therefore the learner could not catch on that it is the present which is being ostensively defined since it could not be contrasted with what is not present.

[2] *A* qualifies as being red, since it is trivially true that every object has the same colour as itself.

? define This obstensively

chooses to indicate at the given moment'. Another way of formulating this ostensive definition is, 'Any event which *is* simultaneous with this is present.'[1] Since this is an ostensive definition it must be temporally unrestricted in its scope; therefore it entails that if at any time an event is *not* simultaneous with *A* then at that time it is *not* present, in which '*A*' is a proper name for the event denoted by the use of 'this' in the ostensive definition of the present.[2] This entails a contradiction. Consider some event *B*, which is not simultaneous with *A*. At the time at which *B* happens it is not simultaneous with *A*, and therefore, according to the negative formulation of the principle of temporal unrestrictedness, at that time is not present. But when an event happens it is by definition present. Therefore, when *B* is present (happens) it is not present. A similar contradiction follows from Ayer's ostensive definition of the past as the class of events which are earlier than the present, which can be reformulated as, 'Any event which is earlier than this is past.' Once again give the proper name '*A*' to the event denoted by 'this' and let *B* be an event later than *A*. Consider a time later than that at which *B* happens. At this time *B* is not earlier than *A*, and therefore at this time *B* is not past. But by definition *B* is past at this time since it is earlier than this time. Therefore at a time at which *B* is past it is not past.

Incidentally, the same sort of absurdities accrue to any attempt to ostensively define an indexical term. For example, suppose that *here* is ostensively defined as follows: 'Any object which occupies this place is here.' Let '*A*' be the proper name of the place denoted by the use of 'this' in the giving of the ostensive definition. This ostensive definition entails that if at any time some object does not occupy the same place as *A*, it is not here. Suppose that I do not occupy place *A*. It would follow that I am not here—that I do not occupy the place I occupy.[3]

The impossibility of ostensively defining *A*-determinations and other indexical terms is due to the fact that they are not general terms. They are strange singular terms since, unlike non-indexical

[1] It is necessary to italicize 'is', for if it is taken as being in the present tense—'is now'—the definition would be circular, 'now' being a synonym for 'present'.

[2] *A* is present, since it is trivially true that every event is simultaneous with itself.

[3] Not all uses of the sentence, 'I am not here', make a contradictory statement. For instance, when a slippery executive says to his secretary, 'I am not here', it means, 'Tell him I am not in.'

singular terms, they admit neither of being ostensively named nor of being ostensively identified. An individual is ostensively named in the course of a baptism or other naming ceremony by being pointed at by the person in authority as he says, 'This is M.M.' or 'This is to be called "M.M." hereafter.' The latter formula is preferable since it makes explicit that M.M. is being ostensively named. If we should attempt to ostensively name the present by saying, 'The class of event simultaneous with this is the present or is to be called "the present" hereafter', the absurd result would follow that we would be required to call this class of simultaneous events 'the present' at moments of time in the future when this class of events is no longer present. And, needless to say, this is not how the expression 'the present' is ordinarily used.[1]

The sentence, 'This is M.M.', can be used, not to bestow a name on M.M., but to ostensively identify M.M., i.e. to demonstratively pick out M.M. as being the same as, or identical with, some described, named or mentioned individual. Could we ostensively identify the present by saying that it is the class of events simultaneous with this? Ordinarily, when I ostensively identify M.M. by saying, as I point to M.M., 'This is M.M.', I am identifying an individual that could be identical with the individual referred to by some previous or subsequent use of the singular term 'M.M.' For example, when I answer the question, 'Who is M.M.?', by pointing to M.M. and saying, 'This is M.M.', I am demonstratively picking out some individual who is identical with the one referred to by the use of 'M.M.' in the preceding question. But when I ostensively identify the present, the class of events which I pick out logically could not be the same as the class of events referred to by some previous or subsequent use of the singular expression 'the present'. Thus, we could not answer the question, 'What is the present?', by saying, 'The present is the class of events which *are* simultaneous with this', since the class of events which we pick out in our ostensive definition could not be identical with the class of events referred to by the prior use of 'the present'

[1] Indexical terms are used in ostensively naming an individual. For example, we can bestow a name on a person by saying, 'This (The person I am now pointing at, The person over there) is to be called "M.M."', or on a place by saying, 'Here shall hereafter be called "Rocky Bottom Creek".' Since indexical terms are instruments used in naming something they themselves cannot be ostensively named, thus the absurdity of 'This is this' or 'Here is here'.

which occurs in the question.[1] 'Now' is not the name of a time because a name can be used at two different times and refer to the same individual.

The upshot of the above discussion is that *A*-determinations, along with other indexical terms, cannot be ostensively defined, named or identified. This brings out that they are neither general terms, names nor descriptions.

[1] Because indexical terms are instruments employed in ostensively identifying an individual, they cannot themselves be ostensively identified. For example, we answer the question, 'Is it now T?', in which 'T' is a (non-indexical) common or proper name for a time by saying, 'It is now T'. But there is no sense to the question, 'Is it now now?', because there is no sense to its answer, 'It is now now'. See L. Wittgenstein, *The Brown Book* (Oxford, 1960), p. 108.

V

THE NON-SENSIBILITY OF *A*-DETERMINATIONS

THE import of the preceding chapter was almost wholly negative, the main point being that because an *A*-statement cannot be rendered by a *B*-statement without loss of factual or informative content *A*-determinations cannot be reduced to *B*-relations. This refutes, in addition to tenet (1) of the *B*-Theory, the major tenet of the Either-Way-Will-Work Theory, which holds that every fact that can be described through the use of an *A*-sentence can also be described through the use of a *B*-sentence, and vice versa. That *A*-expressions cannot be eliminated from a language without lessening its capacity to convey factual information suggests that *A*-expressions have an informative role—that of indicating an 'attribute' or 'property' of an event. Before attempting to reduce *B*-relations to *A*-determinations, I would like to investigate a peculiar feature of the kind of information conveyed by the use of *A*-expressions. What will emerge is that *A*-expressions are not used to designate a *sensible* property or attribute, i.e. a property which is perceived through one or more of the senses. This fact may explain why so many philosophers have been tempted into the error of holding that there is no loss in *factual* content when an *A*-statement is translated by a *B*-statement. These philosophers, following a long empirical tradition going back at least to Berkeley and Hume, have restricted the factual or informative content of a statement to its *empirical* content, which in turn is restricted to its *sensible* content. Since *A*-determinations are not part of the sensible content of a statement they add nothing to its empirical, factual or informative content.

The non-sensibility of *A*-determination can be established by diverse means. One way is to argue that sensible properties are signified by general words, such as 'round', 'red', 'smooth', 'pungent', etc. Since, as seen in the preceding chapter's discussion of the impossibility of ostensively defining *A*-expressions without contradiction, *A*-expressions are not general terms, they do not designate a sensible property. That *A*-expressions are not general does not itself establish that they lack sensible meaning, for there are singular terms, i.e. terms which can have no more than one referent, which do have sensible meaning. Such singular terms are either *definite descriptions* containing at least one general word, such as 'the man who led the Israelites out of Egypt', or a *proper name*, such as 'Moses', which can be replaced or backed up by such definite descriptions. However, *A*-expressions are not definite descriptions or proper names. Unlike proper names, they cannot be used in ostensively naming or identifying an individual, and, unlike definite descriptions, they cannot be used to make ostensive identifications,[1] for the reasons given on pages 66–67. *A*-expressions, like all indexical terms, are peculiar singular terms, since they designate without describing. It was for this reason that Russell called them 'logically proper names', but it would be more accurate to call them 'logically improper names'.

There are different and more intuitive considerations that can be marshalled to show that *A*-expressions are devoid of sensible or empirical meaning. These considerations are similar to those which prompted Kant and the logical positivists to deny that existence is a concept or predicate. When we add existence to our conception of a dollar bill, our concept seems to remain unaltered. It also seems to remain unaltered when we add to our conception the present (past, future) existence of this dollar. Similar considerations apply to our addition of an *A*-determination to our conception of an event or state of affairs. Conceive of Leonard Warren's dropping dead on the stage of the Metropolitan Opera House during a performance of *la Forza del Destino*. Furthermore,

[1] If you should ask, 'Who was the first man to scale Mt. Everest?', I could answer your question by ostensively identifying the individual through the use of the sentence, 'This is the first man to scale Mt. Everest'. Note that your and my use of the definite description, 'the first man to scale Mt. Everest', refer to the same individual. See page 67.

conceive of this event as *occurring* on March 4, 1960. Herein a genuine addition is made to your conception, since you now conceive of this event being 1960 years later than the birth of Jesus. Now add to your conception of Leonard Warren's death on March 4, 1960, that this event has the *A*-determination of presentness (pastness, futurity). By conceiving of it as occurring now (having occurred, occurring in the future) you are not adding any new empirical content to your original tenseless conception: you are not conceiving of the stage on which he fell as being any more solid, or the orchestra as having any more musicians, or Warren's costume as being of a different colour than in your tenseless conception. Since the addition of an *A*-determination does not add to the empirical content of your conception it follows that the *A*-statement, 'Leonard Warren is now dying on March 4, 1960', can be translated without loss of meaning into the *B*-statement, 'Leonard Warren *dies* on March 4, 1960'.

These above considerations are hardly decisive, since it is not clear just what counts as an addition of new empirical meaning to a concept: conceiving seems to be confounded with imagining. All that the above thought-experiments show is that we do not picture something different in our imagination when we add to our imaginative picture of an event that it has a certain *A*-determination. This would seem to show that *A*-determinations are nonsensible. There is a clearer way of showing that *A*-determinations are not sensible properties, which is based on a difference in the logical behaviour of *A*-expressions and words for sensible properties. An object or event can now have a sensible property or a property the criterion for which makes reference to sensible properties only if the object or event now exists or happens, respectively. But an object or event can now have the *A*-determination of pastness (futurity) although it does not now exist or happen. What a strange property it is which an object or event can now have even though it does not exist or happen now.

I shall now elaborate on this surprising feature of *A*-determinations by developing an argument to show both that 'exists (existed, will exist)' is not a predicate of things or continuants and that 'is present (past, future)' is not a predicate of events or states of affairs. I shall confine my remarks to statements having a singular referring expression as their subject. My argument requires that we accept as a premiss that Strawson's account of referring correctly depicts the working of statements containing a singular referring expression as their subject.[1] According to the Strawsonian account, a subject-predicate sentence can be used to make a statement, that is, to say something that can be said to be true or false, only if its subject expression refers successfully. Thus the statement 'The S is ϕ' presupposes the existential statement, 'There is one and only one thing which is S,' such that if this latter statement is false nothing true or false has been said, but rather there has been what Strawson calls a spurious or incorrect use made of the sentence 'The S is ϕ'.

I shall employ certain non-standard terminological distinctions. A distinction is made between *grammatical* and *logical* subject–predicate sentences. A grammatical subject–predicate sentence contains a grammatical subject and predicate expression, the criterion for distinguishing between them being that the predicate contains an assertive or proportional expression, e.g. 'is a . . .'. A logical subject–predicate sentence is a grammatical subject–predicate sentence whose grammatical subject is controlled by a Strawsonian referring rule: therefore, a use of a logical subject–predicate sentence makes a statement only if its grammatical subject refers successfully. The grammatical subject (predicate) of a logical subject–predicate sentence is called the logical subject (predicate) of the sentence. Thus the logical subject is not a non-linguistic entity, as it is in some accounts, but merely a grammatical subject *qua* playing a certain role, viz. being controlled by a referring rule.[2] A statement is said to be either a logical or grammatical subject–predicate *statement* if it is made through the use of either a logical or grammatical subject–predicate sentence respectively. When it is said that 'exists' and 'is present' are not predicates it means that they are not logical predicates, since they are permissible predicates only in purely grammatical subject–predicate sentences.

A vital question, which has been overlooked in previous accounts of referring, concerns the tense of the presupposed existen-

[1] P. F. Strawson, 'On Referring', in *Essays in Conceptual Analysis*, edited by A. Flew (London, 1956).

[2] It might be preferable to define a logical subject as a referring expression in a sentence which is subject to a referring rule. This does not require that a logical subject also be a grammatical subject. Since I shall work, for ease of exposition, exclusively with grammatical subject–predicate sentences, I shall not bother to define a logical subject in this way.

tial statement and how it is related to the tense of the statement presupposing it. My thesis is that a logical subject–predicate statement presupposes an existential statement having the same tense. The statements, 'The *S* is now (was, will be) ϕ' and 'M.M. is now (was, will be) ϕ', presuppose the existential statements, 'There exists now (existed, will exist) one and only one thing which is an *S*' and 'There exists now (existed, will exist) someone named "M.M."', respectively. This thesis shall be developed first for statements whose subject expression refers to a thing for the purpose of showing that 'exists (existed, will exist)' is not a *logical* predicate, and will then be developed for statements whose subject expression refers to an event for the purpose of showing that 'is present (past, future)' is not a *logical* predicate.

The following is a list of *A*-statements whose subject is a thing-expression, along with their presupposed existential statement:

(1) Charlie's only aunt is now bald.
presupposes
(1′) There now exists one and only one aunt of Charlie.
(2) Caesar was assassinated.
presupposes
(2′) There existed a man named 'Caesar'.
(3) The first man to land on the moon will be an Eskimo.
presupposes
(3′) There will exist one and only one first man to land on the moon.

In each case it will be noted that there is a correspondence in tense between the logical subject–predicate statement and its existential presupposition. In saying that (1) is false (true) we are presupposing the truth of (1′), and not just the truth of a tenseless existential presupposition that there *exists* (at some time or other) one and only one aunt of Charlie. The latter is implied by, but does not imply, (1′). Similar considerations hold *mutatis mutandis* for (2) and (3).

Subject–predicate statements in which the subject expression contains an *A*-expression which clashes with the main verb or copula, such as, 'The man who will be the next President is now in the next room', are not counter-examples to my thesis. This statement does presuppose that there now exists one and only one man who will become the next President. Often there is an ambiguity in the use of a descriptive referring expression. When it is said, 'The man who has ϕ was *R*', does it mean that this man had the property ϕ at the time of the reported event or state of affairs—his being *R*—or only that the man who had been or was to be ϕ was *R*? The context in which a statement is made usually resolves such ambiguity. 'The strongest man in the world was born in Texas' could mean that the baby being born was at the time of his birth the strongest man in the world. But obviously the speaker did not intend us to interpret him as asserting this, for it is common knowledge that new-born babes are approximately seven pound weaklings. Thus his statement does not presuppose that there existed at the time of the reported birth a strongest man in the world, only that there existed at that time a man who was to become the strongest man in the world. Therefore, what he said is elliptical for 'The man who was to become the strongest man in the world was born in Texas'.

My above thesis, as well as the more general thesis of Strawson, must be restricted in its application to statements whose predicate does not contain an intentional term, such as 'Santa Claus is now loved by boys and girls'. Obviously this intentional sentence is merely a grammatical subject–predicate sentence (according to my terminology) since its grammatical subject is not controlled by a referring rule. To utter this sentence when its grammatical subject fails to refer successfully is still to say something that is true or false. My thesis thus restricted reads: a logical subject–predicate statement whose predicate is non-intentional presupposes an existential statement of the same tense.

There still remain some few counter-examples to this restricted thesis, but they are not particularly troublesome because they admit of a non-controversial paraphrase which does obey this thesis. For instance, 'Stalin is now influencing Russian politics', certainly does not presuppose that there exists now a man named 'Stalin'. This is a causal statement, 'influencing' being a causal verb. Since a thing or continuant is not a cause, a cause being something that a thing does or that is done with the thing, its subject expression is elliptical for 'Stalin's deeds or actions'. But the subject needs to be made more explicit; in this case it means 'Stalin's past deeds' so that the statement is, 'Stalin's past deeds are now influencing Russian politics', which does presuppose an existential statement of the same tense—'There are now past deeds of Stalin.' 'Dodos are now extinct' is another seeming counter-example, but obviously this statement admits of a non-controversial paraphrase—'There no longer exist Dodos'—which is not even a grammatical subject–predicate statement.

If my thesis is sound it can be shown that no sentence of the form 'The *S* (M.M.) exists now (existed, will exist)' is a logical subject–predicate sentence. The argument I shall offer is a *reductio ad absurdum* which shows that *if* we treat sentences of the form 'The *S* (M.M.) exists now (existed, will exist)' as both logical subject–predicate sentences and as being used to make a contingent statement, then the absurd conclusion follows that no sentence of this form can ever be used to make a false statement. Since these sentences are not used to make either absolutely or hypothetically necessary statements, it is false that they are logical subject–predicate sentences. To make this more apparent let us apply the general rule (as restricted) correlating the tense of a logical subject–predicate statement with that of its existential presupposition to grammatical subject–predicate statements having a naked 'now exists (existed, will exist)' as their predicate:

(4) Charlie's only aunt now exists.
presupposes
(4′) There now exists one and only one aunt of Charlie.
(5) Caesar existed.
presupposes
(5′) There existed a man named 'Caesar'.
(6) The first man to land on the moon will exist.
presupposes
(6′) There will exist a first man to land on the moon.

What immediately strikes us as odd about the above list is that *if* (4), (5) and (6) are taken to be logical subject–predicate statements then each of them presupposes an existential statement with which it is synonymous, and thus it presupposes itself! But according to the Strawsonian account, the use of sentence (4), for example, makes a statement—says something true or false—only if its existential presupposition (4′) is true. But if (4′) is false then (4) is a non-starter, and therefore is not false. From which it follows that no use of sentence (4) could make a false statement, which is a *reductio ad absurdum* of the claim that it is a logical subject–predicate sentence, assuming, of course, that it is used to make a contingent statement. The reason why (4) is not a logical subject–predicate sentence is because 'now exists' is not a logical predicate. Similar considerations hold *mutatis mutandis* for (5) and (6).

Another way of showing that 'exists' is not a logical predicate is to consider the behaviour of 'does not exist'. If 'exists' is a logical predicate then 'does not exist' should also be a logical predicate. But, if we treat 'Pegasus does not exist' as a logical subject–predicate statement then it presupposes 'There exists an individual named "Pegasus"', which is its contradictory. Thus a use of 'Pegasus does not exist' can make a statement only if it is false! But this is an absurd consequence because 'Pegasus does not exist' is not used to make a statement that is either hypothetically or absolutely contradictory.

I am not denying that there are uses for grammatical subject–predicate sentences containing 'now exists (existed, will exist)' as their grammatical predicate, only that this use is for making a logical subject–predicate statement. For example, 'Caesar existed' might be used to inform someone unversed in history that Caesar is not a fictional or mythical person; 'Lions now exist' can be used to make an instantial claim, which would be more perspicuously represented by 'There now exists a lion'; 'Your brother exists' could have a dramatic reinforcing use, which stresses the fact that you must not forget him. It must be stressed that when I deny that 'now exists' is a logical predicate I am concerned only with its naked occurrence, for obviously this term can be expanded by the addition of 'in (at, on) some place (time)' so as to form a logical predicate. Thus 'Caesar existed in 66 B.C.' is a logical subject-predicate statement, which presupposes that a man named 'Caesar' existed, but not that he existed in 66 B.C.

An objection might be raised to my argument to show that 'exists' is not a logical predicate to the effect that subject–predicate sentences containing 'exists' as their grammatical predicate are non-extensional because the 'reference' of the grammatical subject is what would be its 'sense' in an extensional context: 'exists' is an intentional predicate, which causes this shift from reference to sense in the subject expression. Thus 'Caesar existed' is analysed into its formal mode equivalent 'The concept of Caesar had an instance', the latter presupposing only that there *is* a concept of Caesar.[1] This is an interesting analysis but raises certain difficulties. First, it is not too clear what the sense of an expression is and how in some contexts it can become the reference of this expression. Second, this analysis assumes that there are individual concepts but it is very difficult to articulate the relation between a proper name and a set of definite descriptions. Third, it is odd to speak of 'Caesar existed' as non-extensional since it has none of the familiar features of these contexts: the principle of substitutibility holds (If Caesar is the man who conquered Gaul then the man who conquered Gaul existed); and there is no intentional non-existence (If it is true that Caesar existed then there existed a man named 'Caesar'). Even if these three difficulties could be resolved, it would not show that this analysis of 'exists' as an intentional predicate refuted my thesis that 'exists' is not a logical predicate of a *thing*; for according to this analysis 'exists' is a predicate of a *concept* of a thing, or, more accurately, it is 'has an instance' that is a predicate of a concept of a thing.

If my argument to show that 'exists' is not a logical predicate of a thing is sound then it easily can be shown that no sentence predicating 'is present (past, future)' or 'is occurring, happening, taking place (and their corresponding tenses)' of an event or state of affairs is a logical subject–predicate sentence. Aside from philosophical contexts, we almost never say something like, 'My breakfast is past and my dinner is future'; and it is my purpose to show why. First it must be shown that the general rule correlating the tense of a logical subject–predicate statement with that of its existential presupposition holds for statements having an event-expression, rather than a thing expression, as their grammatical subject. The following list shows that this holds:

(7) The falling of the Brooklyn Bridge is now holding up traffic.
presupposes
(7′) There now is one and only one falling of the Brooklyn Bridge.
(8) The falling of the Brooklyn Bridge was noisy.
presupposes

[1] This sort of analysis is given by Wilfred Sellars in his 'Time and the World Order', *op. cit.*, which is reprinted in his *Science, Perception and Reality*, *op. cit.*

(8′) There was one and only one falling of the Brooklyn Bridge.
(9) The falling of the Brooklyn Bridge will be tragic.
presupposes
(9′) There will be one and only one falling of the Brooklyn Bridge.

Since the general rule correlating the tense of a logical subject–predicate statement with that of its existential presupposition holds for statements containing an event-expression as their subject, the *reductio ad absurdum* argument which was employed above to show that 'exists' is not a logical predicate of a thing can be deployed here to show that 'is present (occurring, happening, taking place, and their corresponding tenses)' is not a logical predicate of an event. If we treat sentences of the form 'The ϕing of S is present (past, future)' or 'The ϕing of S is now occurring (happening, taking place, and their corresponding tenses)' as both being logical subject–predicate sentences and as being used to make a contingent statement, then the absurd conclusion follows that no sentence of this form can ever be used to make a false statement. Since these sentences are not used to make either hypothetically or absolutely necessary statements it is false that they are logical subject–predicate sentences. To make this clear let us consider the following list of statements predicating 'is present', etc., of an event, along with their existential presupposition:

(10) The falling of the Brooklyn Bridge is now present (occurring, happening).
presupposes
(10′) There is now one and only one falling of the Brooklyn Bridge.
(11) The falling of the Brooklyn Bridge is past (occurred, happened).
presupposes
(11′) There was one and only one falling of the Brooklyn Bridge.
(12) The falling of the Brooklyn Bridge is future (will occur, will happen).
presupposes
(12′) There will be one and only one falling of the Brooklyn Bridge.

As was the case with statements containing 'exists' as a predicate, each one of the above 'statements' presupposes an existential statement with which it is synonymous. From which the absurd conclusion follows that the sentences used in (10), (11) and (12) could not be used to make a false statement. We can now see the difference between 'The Brooklyn Bridge is now falling', which is a non-starter only when it is false that there now exists one and only one Brooklyn Bridge, and 'The falling of the Brooklyn Bridge is now present', which is a non-starter not only when it is false that there now exists one and only one Brooklyn Bridge but also when it is false that the Brooklyn Bridge is now falling.

In respect to subject–predicate statements containing an event-expression as their subject it is very important to distinguish between the event *referred to* by the subject expression and the event *reported* by the statement. For example, 'S's ϕing is now P' *refers to* S's ϕing and the event reported is S's ϕing's being P. Failure to make this distinction will lead us to confuse rules of referring, which are contextual requirements for the correct use of a sentence, with rules that specify under what conditions a sentence can be used to make a true or false statement, which rules we have called Truth-condition Rules. Strawson, I believe, failed to make this distinction when he wrote:

> One of the correct uses of the present tense is for the description of states of affairs contemporaneous with the describing of them; the past tense is correctly used to indicate that the situation or event reported is temporally prior to the report. . . . A referring rule lays down a contextual requirement for the correct employment of an expression. But the fact that the contextual requirement is satisfied is not a part of what is *asserted* by the use of a sentence containing an expression governed (in this use) by a referring rule; it is, rather, presupposed by the use of the expression.[1]

When this passage is coupled with Strawson's claim that a use of a sentence which fails to satisfy a relevant rule of referring is not false, but rather an incorrect or spurious use of the sentence, the absurd conclusion follows that no tensed sentence can be used to make a false statement.

To save himself from this absurdity Strawson must apply his

[1] Strawson, *Introduction to Logical Theory* (London, 1952), p. 213.

referring rules only to the event referred to by the subject expression (assuming the subject is an event-expression) and not to the event reported (or described) by the statement containing this subject expression. For instance, 'The falling of the Brooklyn Bridge was tragic' does presuppose that the falling of the Brooklyn Bridge is earlier than the making of the preceeding statement. But it certainly does not presuppose that the falling of the Brooklyn Bridge's being tragic is earlier than the making of this statement. If it did, it would presuppose itself. The rule, 'A present (past, future) tense sentence is used to report an event simultaneous with (earlier than, later than) the reporting of it,' is a Truth-condition Rule for the Use of an *A*-sentence: it is not a referring rule.

Immediately after the above quoted passage, Strawson argued that when we make a prediction it is presupposed that the predicted event has not already occurred.

> Thus, if someone says, 'He will die in the course of the next two months', it is linguistically outrageous to reply, 'No, he won't' and then give as one's reason 'He's dead already'.[1]

I believe that this is true, but it is irrelevant to the truth of Strawson's claim that it is presupposed that the event reported—his death in the course of the next two months—is later than the making of the prediction. When we make a prediction it is presupposed that the event predicted is not earlier than the prediction, but it is not presupposed that the event is later than the prediction, for in that case the prediction would presuppose itself.

Once again, in denying that sentences containing 'is present', etc., as their grammatical predicate can be used to make a logical subject–predicate statement, it is not being denied that they have other uses. For example, 'A miracle occurred' could be used to make an instantial claim, which would be more perspicuously represented by 'There was a miracle'; 'The error happened' could have a dramatic reinforcing use which stresses that it is over and done with and that there is nothing we can do about it now. When 'occur', 'happen' and 'take place' are expanded into 'occur in (on at) some place (time, occasion)' or 'just occurred' they can be logical predicates of events, e.g. 'A Presidential election occurred in 1954', 'The wedding just occurred'. The latter are logical

[1] Strawson, *Introduction to Logical Theory* (London, 1952), p. 213.

subject–predicate statements. Surprisingly, 'is present (past, future)' cannot be expanded this way so as to become logical predicates of events or states of affairs. The reason for this is that only a thing, and usually only a special sort of a thing—a person—can be said to be present in some place or to be present at some occasion or on some date. Smith can be present at the Olympics and can be present in the stadium, but an event, such as the 100-yard dash can be neither, though it can happen or take place in the stadium and happen on a certain day. Smith will be present (was present) on the 8th of December makes sense, but not the 100-yard dash will be present (was present) on the 8th of December.

Almost every philosopher who has ever written on time has used as his paradigm for *A*-statements statements of the form, 'The ϕing of *S* is present (past, future).' This way of speaking usually is not vicious, since it easily can be translated into ordinary language by one of these two techniques: '*S* is (was, will be) ϕing' and 'There now is (was, will be) one and only ϕing of *S*'. There is a notational convenience to using sentences of the form, 'The ϕing of *S* is present', since it is easy to compare *B*-statements of the form 'The ϕing of *S* *is* earlier than the ϕing of *R*' with them. The latter is a relational statement, taking event-expressions as its arguments. The problem, as seen in Chapters I and II, is whether events which stand in this *B*-relation must also have *A*-determinations and moreover whether the *B*-relations between events could be reduced to their *A*-determinations. In dealing with these questions it is natural to retain these event expressions in the *A*-statements into which this *B*-statement is analysed and to put in temporal adjectives with tensed copulae after these expressions. The graphic convenience of this notation when dealing with the reduction of *B*-statements to *A*-statements will become obvious in the next chapter when this will be attempted. It is important that whenever this artificial notation is used in a philosophical argument, it be translated into ordinary language in one of the two above fashions so as to make sure that the philosophical thesis which is proven does not depend on the artificiality of this way of speaking. This we shall do when the occasion requires it.

My preceding argument to show that 'exists' and 'is present' are not logical predicates of things and events respectively rests on the premiss that Strawson's account of referring is correct: and to question this premiss is to call into doubt everything that has been said. There are counter-examples to Strawson's account, which consist of situations in which we want to say that an utterance whose subject expression does not refer successfully is false, and not just a non-starter. An example of such a situation might be one in which we would want to say of a person who uses a sentence that fails to refer successfully that he has lied: in order to lie it is necessary that a false statement be made. In such a situation the Russellian theory of descriptions is preferable because it is closer to the ordinary use of 'false'. Even if we grant that the Strawsonian account fails to hold in this type of situation it still leaves my argument untouched. The reason is that there still remain innumerable situations for which Strawson's account holds, and in these situations there is a radical difference between sentences containing an *A*-expression as their predicate and those containing a sensible term as their predicate, in that only the latter are controlled by a Strawsonian referring rule. These situations are sufficient to show a significant difference in the logic of *A*-expressions and sensible terms.

One might feel, nevertheless, that my argument places too much weight on Strawson's mooted theory of referring. There are other theories of referring, such as Russell's, which many philosophers would claim is just as plausible as Strawson's. If there really is a crucial difference in the logic of *A*-expressions and sensible terms, it also should reveal itself within the Russellian theory. I believe that it does.

The main difference between the Russellian and Strawsonian accounts of referring is that, whereas Strawson says that a subject–predicate statement *presupposes* an existential statement, Russell includes this existential statement as part of the *content of the statement*, which in logical form is really an existential statement rather than a subject–predicate statement, as it is for Strawson. Strawson holds that the statement, 'The S is ϕ', presupposes, 'There exists one and only one S-thing', while Russell claims that the logical form of the statement is, 'There exists one and only one S-thing and it is ϕ'. Assuming the correctness of my amendation of Strawson's account to the effect that there is an agreement in tense between the subject–predicate statement and its existential presupposition, then it should be the case that the existential statement that Russell holds to be part of the logical content of the statement should also agree in tense with the original subject–

predicate statement, since the only difference between these accounts is that one holds the existential statement to be presupposed whereas the other holds it to be part of what is stated. Whereas Strawson's amended account of referring holds that 'The *S* is now (was, will be) ϕ' presupposes 'There exists now (existed, will exist) one and only one *S*-thing', Russell's similarly amended account would hold that the logical form of this grammatical subject–predicate statement is, 'There now exists (existed, will exist) one and only one *S*-thing and it is ϕ.'

This amended version of Russell's account of referring can also be utilized to show a crucial difference in logic between *A*-expressions and sensible terms. Absurdities will result from the application of the amended Russellian account to subject–predicate statements containing a naked *A*-expression as their predicate that are analogues of those which result from the application of the Strawsonian account. According to Russell, 'The Brooklyn Bridge now exists', is analysed into, 'There exists now one and only one Brooklyn Bridge and it exists now'. The final conjunct is redundant: on the Strawsonian account the analogue of this absurdity is that the subject–predicate statement presupposes itself, since it presupposes that there exists now one and only one Brooklyn Bridge. 'The Brooklyn Bridge is now red', when analysed in the Russellian manner into, 'There exists now one and only one Brooklyn Bridge and it is red', does not contain a redundant final conjunct. This shows that *A*-expressions are not sensible terms, for only *A*-expressions turn out to be redundant predicates on the Russellian account. An even more startling absurdity results when a naked *A*-expression occurs as the predicate of a negative statement, since for Russell, 'The *S* does not exist now', turns into the contradiction, 'There exists now one and only one *S*-thing and it does not exist now'. The Strawsonian analogue to this absurdity is that this subject–predicate statement presupposes a statement which is its contradictory.

One might challenge my claim that the Russellian existential statement must agree in the tense of its operator with that of the original subject–predicate statement. It is not implausible to hold, as Russell and following him Quine and White did, that the existential operator is tenseless in all contexts, regardless of what is being quantified over. Therefore, 'The ϕing of *S* is now (was, will be) alpha', is to be analysed as, 'There *is* one and only one ϕing

— No, presupposes only that there is such a thing as a pack of wolves & that that particular pack does exist.

of *S* and it is now (was, will be) alpha'. The tensed indication is made not in the operator but in the final conjunct. If this treatment of the existential operator is adopted, then *A*-expressions are not redundant predicates, for 'The ϕing of *S* is now present' turns into 'There *is* one and only one ϕing of *S* and it is now present'. The tense of the final conjunct adds a new item of information to the content of the statement, since the operator is tenseless.

This plausible move is, in my opinion, mistaken. When we quantify over spatio-temporal entities, such as things and events, it is possible to replace the tenseless existential operator with a disjunction of tensed operators as follows: 'There *is* one and only one ϕing of *S* and it is now alpha' is replaced by 'Either there was one and only one ϕing of *S* and it is now alpha *or* there now is one and only one ϕing of *S* and it is now alpha *or* there will be one and only one ϕing of *S* and it is now alpha'.[1] The latter disjunction is at least logically equivalent to the former statement, and therefore should be substitutible for it in any extensional context without changing the truth-value of the statement made. Let the predicate 'alpha' be 'present', and since this is an extensional context we can make this substitution, so as to yield, 'Either there was one and only one ϕing of *S* and it is now present *or* there now is one and only one ϕing of *S* and it is now present *or* there will be one and only one ϕing of *S* and it is now present'. The first and third disjuncts cancel out, since they are contradictory: it is logically impossible that an event which is now present also be past (future). This leaves us with the second disjunct, 'There now is one and only one ϕing of *S* and it is now present', whose existential operator does agree in tense with that of the original subject–predicate statement, 'The ϕing of *S* is now present'; and as a result its final conjunct is redundant. This bears out my claim that the tense of the existential operator in the Russell analysis must agree with that of the original subject–predicate statement of which it is an analysis.

It has been shown that on both the Strawsonian and Russellian accounts of referring, *A*-expressions differ from sensible terms in that only the latter can occur as a logical or non-redundant predicate. That *A*-expressions are devoid of sensible meaning must not lure us into the false view that they do not have an

[1] If we quantify over abstract entities, it is not possible without absurdity to replace the tenseless operator with a disjunction of tensed operators.

informative function. Those philosophers who identify the sensible content of a statement with its empirical content are often tempted to identify in turn the *factual* or *informative* content of the statement with its empirical content, conceived in this narrow manner. This yields the conclusion that *A*-expressions add nothing to the factual or informative content of a statement, and as a result an *A*-statement can be translated into a *B*-statement without loss of factual content. The error in this argument is the identification of the informative or factual content of a statement with its sensible or empirical content. In the ordinary use of 'informative' and 'factual' *A*-expressions do have the function of conveying information about a contingent fact, as for example that *S* is now ϕ, rather than being ϕ in the past or future. A philosopher who identifies the empirical with the sensible cannot be faulted, since 'empirical' is philosophical jargon; however, he is guilty of an error when he then identifies the factual or informative content of a statement with its empirical content, for in doing so he is narrowing or contracting the ordinary use of 'factual' and 'informative'. It probably would help to avoid confusion if the technical term 'empirical' was used as a synonym for 'factual' or 'informative', rather than for 'sensible'.

VI

THE REDUCTION OF *B*-RELATIONS TO *A*-DETERMINATIONS

THAT *A*-determinations cannot be reduced to *B*-relations does not establish that the converse is the case: that the *A*-Series is necessary does not entail that it alone is sufficient. In this chapter tenet (1) of the *A*-Theory, which is also contained in McTaggart's positive thesis, will be defended. Before arguing that the *A*-Series alone is sufficient for accounting for all temporal facts and that therefore the *B*-Series is reducible to the *A*-Series, it will be necessary to critically examine McTaggart's analysis of the *A*-Series. It will turn out that there is a crucial obscurity, hinted at in footnote 1 on page 11, in his concept of the *A*-Series which has gone unnoticed by proponents of both the *A*- and *B*-Theories. Once the concept of the *A*-Series is properly clarified, it will be seen that the *A*-Series, *properly* understood, is sufficient, the *B*-Series being reducible to it; however, this thesis is not nearly as exciting as the one which *A*-Theorists thought they were espousing. While my clarification of the concept of the *A*-Series will tend to trivialize tenet (1) of the *A*-Theory, what it will indicate about our concept of time is far from trivial, and, in fact, is of the highest importance.

Let us now carefully scrutinize McTaggart's positive thesis, which contains his analysis of time. His analysis is phenomenological, being based on the way in which temporal positions *appear* to us, but everything he says could be recast, and for purposes of clarity needs to be recast, in a linguistic idiom which describes the different ways in which we *talk* about temporal positions. He begins by saying:

> Positions in time, as time appears to us *prima facie*, are distinguished in two ways. Each position is Earlier than some and Later than some of the other positions. . . . In the second place, each position is either Past, Present, or Future. The distinctions of the former class are permanent, while those of the latter are not.[1]

This can be rephrased linguistically as:

> There are two fundamentally different ways in which we make temporal determinations. First, we can say that one event *is* earlier (later) than some other event; and, second, we can say that some event is now past (present, future). The sentences used in making claims of the first sort, unlike those used in making claims of the second sort, (logically) cannot be used at two different times to make statements differing in truth-value.

These two different ways of experiencing or talking about time are supposed to determine two different time-series.

> For the sake of brevity I shall give the name of the *A*-Series to that series of positions which runs from the far past through the near past to the present, and then from the present through the near future to the far future, or conversely. The series of positions which runs from earlier to later, or conversely, I shall call the *B*-Series.[2]

McTaggart's concept of the *B*-Series is clear because we can see the connection between it and our use of the expression 'earlier (later) than'. This connection consists in the fact that *earlier* (*later*) *than* is the generating relation of the *B*-Series; that is, given any two non-simultaneous events *x* and *y*, either *x is* earlier (later) than *y* or *y is* earlier (later) than *x*.

Unfortunately, McTaggart's concept of the *A*-Series is not as clear, for while we can see the connection between the *B*-Series and earlier (later) than talk we do not see the connection between the *A*-Series and our use of the predicates 'is present', 'is past' and 'is future'. Since the *A*-Series is a series it must have a generating

[1] McTaggart, *Nature of Existence*, V, 2, *op. cit.*, pp. 9–10.
[2] *Ibid.*, p. 10.

relation. But what could it be?[1] We know only that the generating relation of the *A*-Series, whatever it might be, must contain an *A*-expression, since there is a different *A*-Series at any two different moments of time. The *A*-Series which McTaggart describes cannot be generated by the unqualified *A*-expressions 'past', 'present' and 'future' since the *A*-Series involves not only that the events comprising it are past, present and future but also that they are past and future by varying degrees. The necessity for the latter is contained in McTaggart's description of the *A*-Series as one 'which runs from the *far* past through the *near* past . . .'. The use of the unqualified predicate 'is now past' cannot distinguish between events in the far and in the near past: it determines a 'blob past'. Past, present and future, rather than being the generating relation of the *A*-Series, merely indicate the kind of elements in the *A*-Series, just as +, — and o do not generate the series of integers but only indicate the kind of elements that belong to the series. Before attempting to answer the question as to what could be the generating relation of the *A*-Series, we shall consider McTaggart's arguments for the necessity and fundamentality of the *A*-Series.

McTaggart advances two arguments for the necessity of the *A*-Series, the first showing that change requires an *A*-Series and the second, which is also an argument for the fundamentality of the *A*-Series over the *B*-Series, showing that 'earlier (later) than' can be defined in terms of 'past', 'present' and 'future'. His first argument has as its basic premiss that there cannot be time without change, which would be granted by proponents of both the *A*- and *B*-Theories. The purpose of this argument is to show that there could not be change (and therefore time) without the *A*-Series. The argument, as pointed out on page 10, proceeds by examining every possible candidate for the title of 'change' other than changes in the *A*-determinations of events and showing that none of them is logically possible, and, thus, by a process of elimination, it is inferred that the only change possible is a change

[1] Surprisingly, none of McTaggart's commentators and critics have asked this basic question. The explanation of this strange oversight might be due to the ease with which we can *picture* the *A*-Series, and often this picture involves a correlation between the events of the *A*-Series and the series of integers or the marks on the edge of a ruler. Having such a picture of the *A*-Series we then think that we understand it and forget to ask what is its generating relation. This is a good example of the danger of pictorial thinking.

in an event's *A*-determination. But the only candidates for change which he considers are changes in *events*, and he has no difficulty in showing that such changes are logically impossible. It is logically impossible for any feature of the *B*-Series to change: there can be no increase or decrease in its membership, no change in the qualities of any of its member events, and no change in the *B*-relations between these members. The death of Queen Anne, for example, is determined by the law of identity to remain forever just the kind of event it is: it remains just as sweet, young and innocent as it always was. Neither can it change in any of its *B*-relations to other events: it cannot through diligence and hard work sneak up a few years on the death of Stalin, nor can the death of Queen Anne break out of the escape-proof *B*-Series, since *B*-relations are permanent. From this McTaggart concludes that the *B*-Series alone cannot give us change; and therefore the *B*-Series is a temporal series only because its members have *A*-determinations and change in respect to them, this being the only possible type of change.

A proponent of the *B*-Theory, as pointed out on page 23, would claim that McTaggart has overlooked one possible candidate for the title of 'change', namely that it is *things*, rather than events, that change, and such change consists in a thing having a different quality at one phase of its history than it has at an earlier or later phase; e.g. a poker changes in that it is at one time hot and at a later time cool. Such qualitative changes *in* things, which were called changes *in* time on page 22, can be analysed solely in terms of the *B*-relations of subsequence and precedence between qualitatively different states which comprise the history of a single thing. Moreover, these qualitatively different states of a single thing can be described by *B*-statements; e.g. 'The poker *is* hot on Monday' and 'The poker *is* not hot at times other than Monday' McTaggart raised the following objection to this analysis:

> This makes no change in the qualities of the poker. It is always a quality of that poker that it is one which is hot on that particular Monday. And it is always a quality of that poker that it is one which is not hot at any other time. Both these qualities are true of it at any time—the time when it is hot and the time when it is cold. And therefore it seems to be erroneous to say that there is any change in the poker.[1]

[1] McTaggart, *op. cit.*, p. 15.

McTaggart's reply fails because of an equivocation on the word 'always'. He argues that because the tenseless statement 'The poker *is* hot on Monday' is *always* true, i.e. true independently of time in the sense that the use of this sentence makes a true statement every time it is used, the state of affairs described by this statement must *always*, i.e. in the temporal sense of 'always', be occurring or existent. Because statements describing events in a tenseless manner are *always* true does not require that these events are sempiternal: to claim the opposite is an unwarranted addition of an eternalistic ontology to a tenseless logic.

McTaggart's other argument to show the dependency of the *B*-Series upon the *A*-Series fares no better than his first one. This argument, which is never stated explicitly, is that our understanding of the concept of earlier (later) than essentially involves reference to the concepts of past, present and future. *Earlier than* is a temporal relation only because its relata have *A*-determinations. At one place, he claimed that we could *define* 'earlier than' in terms of the *A*-determinations of present, past and future:

> The term *P* is earlier than the term *Q*, if it is ever past while *Q* is present, or present while *Q* is future.[1]

If it is possible to define 'earlier than' in terms of 'past, present and future' it will follow not only that there could be no *B*-Series without an *A*-Series but also that the *B*-Series is reducible to the *A*-Series, since the generating relation of the *B*-Series is definable in terms of *A*-determinations.

McTaggart's attempt to define the *B*-relation 'earlier than' in terms of the three *A*-determinations fails because of circularity. The *definiens* of the definition contains the statements, '*P is* past while *Q is* present' and '*P is* present while *Q is* future', which mean the same as, respectively, '*P is* past at *Q*' and '*Q is* future at *P*'. The predicates '____ *is* past at ____' and '____ *is* future at ____ are synonyms for '____ *is* earlier than ____' and '____ *is* later than ____' respectively, in that all four are tenseless two-place predicates which, when event expressions not containing *A*-expressions are substituted for their blank spaces, express a timelessly true or false statement about a *B*-relation between two events, i.e. they make *B*-statements. '*Is* past (present, future) at some event or date', unlike 'is now past (present, future)', does not indicate an

[1] McTaggart, *op. cit.*, p. 271.

A-determination; and therefore it cannot be used to determine the position of an event in the *A*-Series. McTaggart is cheating by allowing '*is* past (future) at' to count as *A*-expressions.[1] If the copulae of the statements in the *definiens* are taken to be tensed—'*P* is now past while *Q* is now present' and '*P* is now present while *Q* is now future'—then the *definiendum* is not logically equivalent to the *definiens*; for it might be the case that *P is* earlier than *Q*, but be false that *P* or *Q* is now present, in which case the *definiendum* is true and the *definiens* false.

If we take the *A*-Series to be determined by an unqualified past, present and future—a past, present and future that do not admit of degrees—then it can be shown that the *B*-Series cannot be reduced to the *A*-Series, this being due to the fact that the *B*-relation 'earlier than' cannot be defined in terms of an unqualified 'past, present and future'. In the first place, it is fallacious to argue that *earlier than* can be analysed in terms of *A*-determinations because there are *A*-statements which imply a *B*-statement but no *B*-statements which imply an *A*-statement; for example, '*P* is present and *Q* is future' entails, but is not entailed by, '*P is* earlier than *Q*'.[2] For the fact that a statement containing a term *X* entails, but is not entailed by, a statement containing *Y* does not show that *Y* is analysable in terms of *X*. 'All *animals* are mortal' entails, but is not entailed by, 'All *men* are mortal', but this surely does not show that men are analysable purely in terms of the concept of an animal. Thus, while it is true that there is an asymmetry in informative or factual content between a language composed wholly of *B*-sentences and one composed wholly of *A*-sentences, in that everything sayable in the former is sayable in the latter but not vice versa, it does not establish that *B*-relations are reducible to *A*-determinations, and therefore that the *B*-Series is reducible to the *A*-Series.

Another attempt to reduce the *B*-Series to a *pure A*-Series, i.e. an *A*-Series determined by the use of an unqualified 'past, present

[1] A similar difficulty haunts Wilfrid Sellars' analysis of '*X is* earlier than *Y*' into 'Either *X* is present and *Y* is future *or X* was present and *Y* future (*quo ad X*) *or X* will be present and *Y* future (*quo ad*) *X*' ('Time and the World Order', *op. cit.*, p. 546); for '*Y* future (*quo ad*) *X*' means the same as '*Y is* future at *X*', which in turn means the same as '*Y is* later than *X*'.

[2] I committed this fallacy in my paper, 'Tensed Statements', *The Philosophical Quarterly*, 12 (1962), and am indebted to J. J. C. Smart for pointing this out in his paper, '"Tensed Statements": A Comment', *op. cit.*

and future', which is also unsuccessful, involves giving a 'definition in use' of 'earlier than' in terms of the three unqualified *A*-determinations, taken together. By a *definition in use* of a term is meant one which gives an analysis of a statement containing this term: we shall assume that a necessary, though possibly not sufficient, condition for a successful analysis is that the *analysandum* and the *analysans* are logically equivalent. Let us try to analyse a *B*-statement in terms of a logically equivalent disjunction of *A*-statements:

(I) *P is* earlier than *Q* . ≡ . *P* is past and *Q* is present *or P* is past and *Q* is future *or P* is present and *Q* is future

in which '*P*' and '*Q*' are expressions referring to an event or state of affairs. It is quite apparent that (I) is not a logical equivalence, for whereas it could not be the case that the disjunction of *A*-statements in the *analysans* is true and the *B*-statement in the *analysandum* is false it could be the case that the *analysandum* is true and the *analysans* is false. The reason for this, which has already been given above in the discussion of McTaggart's definition, is that if *P* and *Q* are either both past or both future, *P* being *more past* than *Q* or *Q* being *more future* than *P*, then each of the three disjuncts would be false. To give an adequate definition in use of a *B*-relation in terms of an exclusive disjunction of *A*-statements we must enlarge our concept of an *A*-determination so as to include *more past* and *more future* as well as an unqualified past, present and future. With this expanded concept of an *A*-determination we can add two disjuncts to the *analysans* of (I) as follows:

(II) *P is* earlier than *Q* . ≡ . *P* is past and *Q* is present *or P* is past and *Q* is future *or P* is present and *Q* is future *or P* is more past than *Q or Q* is more future than *P*

Reduction (II) can be applied to an *analysandum* with any number of terms. For example, '*P is* earlier than *Q* and *Q is* earlier than R', by repeated uses of the principle of Distribution and the crossing out of contradictory disjuncts, reduces to, 'Either *P* is past and *Q* is present and R is future *or P* is past and R is more future than *Q or P* is present and R is more future than *Q or P* is more past than *Q* and *Q* is more past than R *or Q* is more future than *P* and R is more future than *Q or P* is more past than *Q* and R is present.' If there are four terms in the *analysandum* there will be eight disjuncts. I haven't had the courage to work it out for more terms, but, in principle, it could be worked out for an *analysandum* describing the entire *B*-Series.

What is to be concluded from this discussion is that the *B*-Series cannot be reduced to a *pure A*-Series but that it can be reduced to an *impure A*-Series, i.e. an *A*-Series which is determined not only by an unqualified past, present and future but also by more past and more future.[1] Thus the thesis of McTaggart and tenet (1) of the *A*-Theory that the *B*-Series is reducible to the *A*-Series is correct, provided that we are speaking of an *impure A*-Series. This qualification greatly trivializes their thesis. Not only can the *B*-relation *earlier than* be analysed in terms of the *five A*-determinations, taken together, but the same is true for the *B*-relation *simultaneous with*. It can be reduced as follows:

(III) *P is* simultaneous with *Q* . ≡ . (*P* is present and *Q* is present) *or* (*P* is past and *Q* is past and it is not the case that either *P* is more past than *Q* or *Q* is more past than *P*) *or* (*P* is future and *Q* is future and it is not the case that either *P* is more future than *Q* or *Q* is more future than *P*)

Once again, it was necessary to use 'more past (future)' in carrying out the reduction.

Notice that a *B*-statement has been reduced by (II) and (III) not to a single *A*-statement but to a disjunction of *A*-statements, which disjunction is not itself an *A*-statement, since it is made through the use of a sentence for which it is the case that it cannot be used at two non-simultaneous times to make statements differing in truth-value. Our discussion in Chapter IV of the impossibility of reducing *A*-determinations to *B*-relations considered only analyses of an *A*-statement into a *single B*-statement. Even if we permit the *analysans* to contain a disjunction of *B*-statements, it is not possible to analyse an *A*-statement into *B*-statements. '*P* is present and *Q* is future' certainly is not logically equivalent to '*P is* earlier than *Q or P is* later than *Q or P is*

[1] It might be pointed out that the concept of more past (future) than is not a metrical concept. Even if time has no intrinsic metric (due to there being no ultimately discrete events) and we have not laid down conventional co-ordinative definitions for the metricization of time, we can say that one temporal interval is more past (future) than another due to the fact that these intervals overlap, one of them containing the other as a part of itself. For an excellent discussion of this see A. Grünbaum, *Philosophical Problems of Space and Time, op. cit.*, Chapter 1.

simultaneous with *Q*', for the latter is a tautology, assuming that '*P*' and '*Q*' refer successfully, while the former is contingent. And even if we make the latter non-tautological by dropping one of its three disjuncts it could still be true and the *A*-statement in the *analysans* be false. No one, fortunately, has ever suggested such an analysis.

In summary: an impure, but not a pure, *A*-Series is also a *B*-Series, in that a statement describing the *A*-determinations of events entails a statement describing tenseless *B*-relations between these events: this is seen by the fact that any one of the disjuncts in the *analysans* of (II) entails the *B*-statement in the *analysandum*. However, a *B*-Series is not also a specific *A*-Series, in that a statement describing a *B*-relation between events does not entail a statement describing the specific *A*-determinations of these events, but only a disjunction of statements describing different possible combinations of the *A*-determinations of these events. Thus, an *A*-Series is already some specific *B*-Series, but a *B*-Series is not at the same time some specific *A*-Series. This conclusion helps us to see why McTaggart's attempt to derive the *B*-Series from a conjunction of the *A*-Series with a *C*-Series, i.e. a series whose generating relation is non-temporal, is either pointless or futile.

> We can now see that the *A*-Series, together with the *C*-Series is sufficient to give us time. . . . The *B*-Series . . . is not ultimate. For given a *C*-Series of permanent relations of terms, which is not in itself temporal, and therefore is not a *B*-Series, and given the further fact that the terms of this *C*-Series also form an *A*-Series, and it results that the terms of the *C*-Series become a *B*-Series, those which are placed first, in the direction from past to future, being earlier than those whose places are *further in the direction of the future*.[1]

If the *A*-Series referred to in this quotation is an impure *A*-Series, then it is already a (specific) *B*-Series; and therefore it is pointless, since unnecessary, to combine this *A*-Series with another series in order to derive the *B*-Series. But if the *A*-Series referred to by McTaggart is a pure *A*-Series then it can be shown that his attempt to derive the *B*-Series in the manner described is futile. Consider the problem of correlating the members of a *pure A*-Series with the members of a *C*-Series, such as the series of inte-

[1] McTaggart, 'The Unreality of Time', *op. cit.*, p. 462. My italics.

gers. There is no problem in correlating the present event(s) in the pure *A*-Series with the number o in the *C*-Series of integers. But it is impossible to correlate the past and future events of this *A*-Series respectively with the minus and plus integers because this *A*-Series contains a 'blob past and future', i.e. one which admits of no distinction between more past and more future. To make this correlation we must import a structural order into the past and future of the *A*-Series by introducing the concepts of more past and more future, which, incidentally, is exactly what McTaggart does when he speaks of terms that are '*further* in the direction of the future'. We must say something like 'the more past (future) an event is in the *A*-Series the smaller (larger) the integer with which it is correlated'.

Having clarified McTaggart's ambiguous concept of the *A*-Series and shown that only the impure *A*-Series is a series, we can now answer our original question concerning the generating relation of the (impure) *A*-Series. The generating relation of *an* *A*-Series is *earlier than*, which is also the generating relation of the *B*-Series. Since a *B*-Series must also be *an* *A*-Series, though not any one specific *A*-Series, it follows that the generating relation of the *B*-Series must also generate *an* *A*-Series, though not any one specific *A*-Series. Concerning the generating relation of some specific *A*-Series, the following answer can be given. The *A*-Series is formed from a conjunction of two different series having a common number, which serves as the *terminus ad quem* of one and the *terminus ab quo* of the other. There is a series whose generating relation is *more past than*, and there is another series whose generating relation is *more future than*. These two series can be combined to form a single (impure) *A*-Series because the *terminus ad quem* of the series of past events is the present event(s) while the *terminus ab quo* of the series of future events is also the present event(s).

The generating relation of *an* *A*-Series, as well as that for *the* *A*-Series, makes use of the concepts of more past and more future. It might be objected, and not without plausibility, that these concepts involve a tenseless *B*-relation, because a statement of the form, '*P* is now more past than *Q*', must be *analysed* into a conjunction of the *B*-statement, '*P is* earlier than *Q*', and the *A*-statement, '*Q* is now past'. And, if this is so, our reduction of *B*-relations to *A*-determinations by reduction (II) is circular, since the final two disjuncts in the *analysans*—the ones using 'more past' and 'more future'—contain *B*-statements. There are two ways of countering this objection. First, it could be claimed that the statement, '*P* is more past (future) than *Q*', is an *A*-statement according to criteria (*A*)[2] and (*A*)[3], for it is made through the use of a sentence which can be used at two different times to make statements differing in truth-value. This, however, is not decisive in showing that this statement cannot be analysed into a conjunction of an *A*- and a *B*-statement; for the conjunction of an *A*- and a *B*-statement will always make an *A*-statement, just as, analogously, the conjunction of a contingent and a necessary statement will always make a contingent statement.

A more effective way to meet this charge of circularity in our reduction of *B*-relations to *A*-determinations is to claim that '*P* is more past than *Q*' *entails* '*P is* earlier than *Q*' but deny that we must *analyse* '*P* is more past than *Q*' into the conjunction of '*P is* earlier than *Q* and *Q* is now past'. If '*P* is more past than *Q*' did not entail '*P is* earlier than *Q*' reduction (II) would fail, it being necessary that any one of the disjuncts in its *analysans* entail the *analysandum*. Since we require for a successful analysis that the *analysans* and *analysandum* be logically equivalent, the *analysans* in a successful analysis will entail the *analysandum*. If this is to be a reason for condemning an analysis as viciously circular then the paradoxical result follows that no analysis can succeed. If the charge of circularity against (II) is to rest on some basis other than the paradox of the impossibility of analysis it must be shown not just that '*P* is more past than *Q*' *entails* '*P is* earlier than *Q*' but that it must be *analysed* in terms of this *B*-statement with the *A*-statement '*Q* is now past'. And I know of no argument that can be given to show that we must accept this as an *analysis* of '*P* is more past than *Q*'.

I am not very happy with this reply to the charge of circularity. It might be thought that this rather indecisive result could be avoided by offering a different reductive analysis of *B*-relations to *A*-determinations than that given in (II). For example, instead of 'more past than' we could have used 'preceded', 'was earlier than' or 'occurred a longer time ago than', etc., but they are just as susceptible to the charge of circularity as is 'more past than'. One might try to escape from the charge of circularity by giving a metalinguistic recasting of (II) as follows:

(IV) *P is* earlier than *Q* . ≡ . *P* is present and *Q* is future *or P* is past and if at the time *P* occurs someone were to use a sentence synonymous with '*Q* is now future' a true statement would be made *or P* is future and if at the time *P* occurs someone were to use a sentence synonymous with '*Q* is now future' a true statement would be made

This metalinguistic way out comes close to working, but it comes to grief in its second and third disjuncts over the use of 'at the time *P* occurs', which is a synonym for the *B*-expression 'simultaneously with *P*'. While this reduction, unlike (II), is above reproach for circularity, it nevertheless fails to analyse the *B*-relation *earlier than* exclusively in terms of *A*-determinations.

The following metric analysis is another attempt to analyse *earlier than* without invoking any *B*-relation:

(V) *P is* earlier than *Q* . ≡ . *P* is past and *Q* is present *or P* is past and *Q* is future *or P* is present and *Q* is future *or P* is *n* number of time-units past and *Q* is n_1 number of time-units past and $n_1 < n$ *or P* is *m* number of time-units future and *Q* is m_1 number of time-units future and $m < m_1$

The symbol '<' in the last two disjuncts means 'is less than', and therefore this analysis is not overtly circular. This reduction works only if it is not necessary to make use of *B*-expressions in explaining what is meant by the metric expressions in the last two disjuncts. Unfortunately it is not possible to explain what is meant by '*P* is *n* number of time-units past' without reference to the *B*-relations of precedence and simultaneity. Reference must be made to some cyclical process by which temporal congruence is defined: this cyclical process involves *successive* events—events which are *earlier* and *later* than each other. Also, that an event occurred *n* number of time-units ago requires that this event be correlated by the relation of *simultaneity* with events that occur in our clock.

It might be thought that the aurora of circularity which surrounds reduction (II) is due to the fact that the *A*-statements in its *analysans* contain the *non-logical* predicates 'is present (past, future)'. As noted on pages 81–82, it is important that these jargonistic sentences be translated into either tensed *logical* subject–predicate sentences or into tensed existential sentences, so as to make sure that the philosophical thesis which is being argued for does not rest on this artificial way of speaking. Possibly this aurora of circularity can be eliminated by carrying out a translation of these sentences into ordinary language in one of the two ways indicated:

(VI) *S*'s being ϕ *is* earlier than R's being ϕ . ≡ . *S* was ϕ and R is now ϕ *or S* was ϕ and R will be ϕ *or S* is now ϕ and R will be ϕ *or S* was ϕ before R was ϕ *or S* will be ϕ before R will be ϕ[1]

(VII) *S*'s being ϕ *is* earlier than R's being ϕ . ≡ . There was a being ϕ of *S* and there is now a being ϕ of R *or* There was a being ϕ of *S* and there will be a being ϕ of R *or* there is now a being ϕ of *S* and there will be a being ϕ of R *or* there was a being ϕ of *S* before there was a being ϕ of R *or* there will be a being ϕ of *S* before there will be a being ϕ of R

The only difference between (VI) and (VII) is that one employs subject–predicate sentences in its *analysans* while the other employs existential statements. Philosophers holding to a thing-ontology will prefer (VI) while those holding to an event-ontology will prefer (VII), but this issue will not concern us here.

Reductions (VI) and (VII), like (II), falter in their final two disjuncts. Whereas our suspicions were aroused by the use of 'more past (future) than' in these disjuncts of (II), our problems with (VI) and (VII) arise from their use of the seemingly *B*-like expression 'before' in their last two disjuncts. Their use of 'before' does not make them overtly circular, since the *relational* expression '*is* earlier than' occurs in the *analysandum*, rather than the *connective* 'before'. 'Before' is a *connective* because only sentences can be its arguments, unlike the *relation* '*is* earlier than' which takes only event-expressions as its arguments. 'Before' is the counterpart in connective discourse to '*is* earlier than' in relational discourse. If the *B*-statement in the *analysandum* of (VI) and (VII) is translated from relational discourse to connective discourse, it becomes, '*S is* ϕ before R *is* ϕ'. The latter is a synonym in connective discourse for the relational statement, '*S*'s being ϕ *is* earlier than R's being ϕ'. One could argue that '*S* was ϕ before R was ϕ' must be *analysed*

[1] A similar translation of the *analysans* of reduction (III) on page 93 would be: '*S* is now ϕ and R is now ϕ *or S* was ϕ when R was ϕ *or S* will be ϕ when R will be ϕ.'

into the conjunction of the *B*-statement '*S is* ϕ before R *is* ϕ' with the *A*-statement 'R was ϕ', thus rendering (VI) circular, since a statement synonymous with the *analysandum* is contained in the *analysans*. Instead of '*S* was ϕ before R was ϕ' we could have used '*S* had been ϕ when R was ϕ', but the use of the compound tense raises the same problem as the use of 'before'; once again it could be claimed that this statement must be *analysed* into '*S is* ϕ before R *is* ϕ and R was ϕ'.

The charge of circularity against (VI) and (VII) is to be met in the same way as was the similar charge against (II), which is their counterpart in relational discourse. Admittedly, '*S* was ϕ before R was ϕ' *entails* '*S is* ϕ before R *is* ϕ', as well as its relational discourse synonym, '*S*'s being ϕ *is* earlier than R's being ϕ'. If it did not, the analysis would not succeed; but this does not require us to *analyse* '*S* was ϕ before R was ϕ' into '*S is* ϕ before R *is* ϕ and R was ϕ'. Once again, I am not very happy with this rather indecisive reply. The need to make use of the expression 'more past (future) than' or 'before' in our reduction of a *B*-statement to a disjunction of *A*-statements somewhat trivializes tenet (1) of the *A*-Theory concerning the reducibility of *B*-relations to *A*-determinations.[1] One feels that the only true-blue *A*-determinations are the unqualified past, present and future: 'more future than' smells very much like a relation, since it looks like a two-place predicate. At any rate, it is a very different sort of *A*-expression than the simple 'is present (past, future)'. While the need to make use of 'more past (future) than' or 'before' wins a rather hollow victory for the *A*-Theory and McTaggart, what it indicates about our concept of time is far from trivial. It shows that our concept of time involves in addition to the three unanalysable concepts of past, present and future the notion of a structural order, and it was because of the latter that we had to introduce the concepts of *more past* or *before*.

Even if reductions (II), (VI) and (VII) are circular, they still are of the first importance. There is no question about their *analysandum* and *analysans* being logically equivalent. The *B*-statement in their *analysandum* entails a disjunction composed of five disjuncts, each of which indicates the *A*-determination of an event; and this

[1] Our reduction in (III) of the *B*-relation *simultaneous with* to *A*-determinations also makes use of 'more past (future) than' in its *analysans*: and if it is true that this expression must be analysed in terms of the *B*-relation of *earlier than*, (III) fails to reduce a *B*-relation to *A*-determinations *exclusively*.

holds regardless of whether the final two disjuncts *in addition* make reference to a *B*-relation. This bears out the intuitions of McTaggart and several *A*-Theorists that there cannot be a *B*-relation of *earlier than* between two events unless these events have *A*-determinations. That events cannot stand in *B*-relations to each other unless they have *A*-determinations is independent of the thesis of Chapter IV that *A*-determinations are ineliminable. Because facts about the *A*-determinations of events cannot be reported except through the use of an *A*-sentence does not establish that these events cannot bear *B*-relations to each other unless they have *A*-determinations. Thus, our reductive analyses of *B*-relations to *A*-determinations, even if circular, are of great importance, showing that there could not be a *B*-Series unless there also is an *A*-Series.

Part Three

LOGICAL ASYMMETRIES BETWEEN THE PAST AND FUTURE

VII

THE IMPOSSIBILITY OF BRINGING ABOUT THE PAST

THE preceding Part explored the logical connections between *A*- and *B*-expressions. It was established that the *A*-determinations of events are both ineliminable and necessary for these events having *B*-relations to each other: and these results hold even if our attempts to reduce *B*-relations to *A*-determinations were unsuccessful. In arguing for these theses many interesting features of the logic of *A*-expressions were uncovered—that *A*-expressions have an informative role but not that of indicating a sensible property, that *A*-expressions are unusual singular terms in that they are neither proper names nor definite descriptions. This Part will explore a different dimension of the logic of *A*-expressions concerning logical asymmetries between the past and the future. It will be shown that many of our fundamental concepts, such as causality, action, deliberation, choice, intention, memory, knowledge, truth, possibility and identification, logically presuppose an asymmetry between the past and future.

It will be recalled that the *A*-Theory held temporal becoming to be an objective property of all events and claimed that because of this the past and future differ ontologically, the future being open and the past closed. Since past events have become present they have already won their ontological diplomas, unlike future events which still exist in a limbo of mere possibility. Our ordinary concept of time contains this view, as witnessed by such gems as, 'Don't cry over spilt milk', 'It's like locking the barn after the horse has escaped', and 'What has been done cannot be undone'. This common-sense distinction between the open future and the closed past can best be clarified by reference to certain logical asymmetries in our ordinary ways of talking about the past and the future: for example, it is meaningful to speak of causing, deliberating, deciding and intending about a future, but not a past, event, while, on the other hand, it makes sense to speak of having a trace (such as a memory, photograph or scar) only of a past event.

Any one of these logical asymmetries, when viewed in isolation from the others, will seem to be a rather unexciting tautology—hardly the sort of a thing we would want to get down on our hands and knees and worship. However, when a synoptic view is taken of them they no longer will appear trivial or unimportant. It will then be seen that these logical asymmetries interlock and mutually support each other like the legs of a tripod, such that one of them cannot be jettisoned without giving up the others as well. Together these asymmetries form the basis of our common-sense conceptual system. These intimately related asymmetries will be used in Chapters VIII and IX to show respectively that: (i) statements about the past are now either true or false while some statements about the future are neither; and (ii) future individuals, unlike past ones, cannot now be identified. These two logically independent asymmetries between the past and future support the claim of tenet (3) of the *A*-Theory that there are significant ontological differences between the past and future.

The *B*-Theory's assessment of these logical asymmetries would be as follows. There are no logical asymmetries between past and future because events are not intrinsically past and future, but only earlier than (past at) and later than (future at) some other chosen event(s). At best, certain of the above logical asymmetries can serve as criteria for distinguishing between earlier and later events relative to some chosen event. In other words, none of these criteria enable us to pick out some event as uniquely present, and therefore other events as past and future; since it is true of any time that only events later than (future at) that time could be deliberated about, etc., and that only events earlier than (past at) that time could be remembered, etc. Not only do these logical asymmetries not serve as criteria for determining what event(s) is uniquely present, but there are compelling reasons, based on the

Linguistic and Psychological Reductions, for believing that the distinctions of past, present and future are 'subjective'. The challenge posed by these Reductions will be met in Part Four, where it will be argued that temporal becoming is objective.

The *B*-Theory, however, cannot accept the two logical asymmetries above—namely, that some statements about the future are neither true nor false, and that no future individual can be identified—even when they are demoted to the status of mere criteria for distinguishing between an earlier and later time. The *B*-Theory cannot accept them because it countenances no ontological distinctions between events occurring at different times. All events comprising history are equally real and determinate, and can be completely described by timelessly true or false *B*-statements. By a parity of reasoning, it should also be true that all individuals, regardless of when they exist, are equally real and determinate, and therefore should be identifiable through the use of suitable referring expressions, such as proper names and definite descriptions.

We shall begin our articulation of the complex logic of the above family of inter-related logical asymmetries between past and future by first considering the concept of causality and the asymmetry between past and future which is built into this concept. The question which philosophers have asked is, 'Could a present event cause a past event?'[1] It is true but unhelpful to reply that this is logically impossible since it is an analytic truth that a present cause cannot have a past effect. The reason why this reply is unhelpful is that it does nothing towards answering the question of whether or not we could alter this analytic truth by suitably changing the rules for the use of the word 'cause' and

[1] The following authors deal with this question: M. Dummett and A. Flew, 'Can an Effect Precede its Cause?' (symposium), *Aristotelian Society Supp.*, Vol. 28 (1954); M. Black, 'Why Cannot an Effect Precede its Cause?' *Analysis*, 16 (1956); A. Flew, 'Effects Before Their Causes? Addenda and Corrigenda', *Analysis*, 16 (1956); M. Scriven, 'Randomness and the Causal Order', *Analysis*, 17 (1956); A. J. Ayer, *The Problem of Knowledge*, *op. cit.*, pp. 170–175; D. F. Pears, 'The Priority of Causes', *Analysis*, 17 (1957); A. Flew, 'Causal Disorder Again', *Analysis*, 17 (1957); R. M. Chisholm and R. Taylor, 'Making Things to Have Happened', *Analysis*, 20 (1960); M. Dummett, 'Bringing about the Past', *Philosophical Review*, XXIII (1964), reprinted in *The Philosophy of Time*, edited by Gale; S. Gorovitz, 'Leaving the Past Alone' *Philosophical Review*, XXIII (1964); and R. M. Gale, 'Why a Cause Cannot Be Later Than Its Effect', *The Review of Metaphysics*, XIX (1965). Also to be consulted in this connection are the articles by C. D. Broad, C. J. Ducasse and A. Flew in *The Philosophy of C. D. Broad*, edited by P. Schilpp, *op. cit.*

still have a recognizable analogue of our present concept of causality so that while it still is required that a cause be uniformly connected with its effect and could be used to bring its effect about, etc., it is no longer required that a present cause cannot have a past effect. This question, then, is not whether there could be a counter-example to the analytic truth that a cause cannot have a past effect, which, of course, is logically impossible, but whether there could be a 'counter-stipulation-example' to this analytic truth, that is, a description of a constant conjunction between events (such as is given in para-normal psychology and science-fiction) which is of such a sort so as to justify our giving up this analytic truth.[1] A counter-stipulation-example usually involves a description of a case in which an agent, in ignorance of the past, performs some intentional action to bring about a past event and later discovers that this event did happen.

The aim of this chapter will be to try and show why there could not be any acceptable counter-stipulation-example to this analytic truth. The overall strategy will be to articulate the conceptual centrality of this analytic truth by showing that a change in it will cause absurdities to break out in certain neighbouring concepts with which the concept of cause has logical liaisons, such as action, possibility, intention, deliberation, memory, experimentation, responsibility and punishment; and when we try to protect ourselves from these absurdities by altering these latter concepts as well, we will have twisted our concept of causality, along with these other concepts, out of recognition. In other words, it will be argued that to give up this analytic truth will require such a drastic overhauling of the basic concepts of our conceptual system that they will not be recognizable analogues of their former self. Obviously, I cannot give any precise criterion for just when an altered concept ceases to be a permissible analogue of its former

[1] The way in which we have posed the question of whether or not we could give up this analytic truth involves acceptance of some form of the conventionalist's theory of necessity; however, the same sort of question could be raised if we operated with a platonic theory of necessity. Since the platonist denies that necessity is dependent in any way on linguistic rules, he would not allow us to speak of denying, or giving up, an analytic truth by changing linguistic rules. In regard to cases where supposed analytic truths were denied, e.g. development of non-Euclidean geometries, he would say that this shows only that they really were not analytic truths but only believed to be such. But, then, we could raise the question of whether the claim that a present event cannot cause a past event really is an analytic truth or is only believed to be one.

self. While there is room for disagreement about these matters, we can reach agreement upon the logical connections between the concept of causality and these other neighbouring concepts, so that we can determine what changes must be made in the latter so as to accommodate a change in the former. Whether we then wish to say that these changes are so extensive that these concepts are not permissible analogues of their former selves is of secondary importance. However, I do believe that it can be shown that the extended concept of causality which permits a present event to cause a past effect requires such a drastic change in neighbouring concepts that there will be little room for disagreement as to whether it is a legitimate analogue to the original concept of causality.

To bring out the conceptual centrality of the analytic truth that a present cause cannot have a past effect, we shall try to devise a counter-stipulation-example to this truth and shall note what corresponding conceptual reforms it requires us to make. But before doing this it will be necessary, first, to make some preliminary remarks about causality, and, second, to lay down certain necessary conditions which must be fulfilled by any successful counter-stipulation-example.

I. PRELIMINARY REMARKS ABOUT CAUSATION

Fortunately, for our purposes it will not be necessary to define or analyse causality, which probably cannot be done; but it will be essential to distinguish, on the one hand, between a proximate and a remote cause, and, on the other, between a cause and a mere condition. A proximate cause is a cause (in some undefined sense) which bears one of the following three temporal relations to its effect: (1) it can be simultaneous with its effect (to be called a 'proximate $cause_1$'); it can be temporally contiguous with its effect (to be called a 'proximate $cause_2$'); or (3) it can be temporally separated from its effect by initiating some process that in the normal course of things terminates in its effect (to be called 'proximate $cause_3$'). The best way to bring out the differences between these three types of proximate cause is to consider some examples of each.

When a lead ball is released on a not too tightly packed sandbag it immediately begins to indent this bag and continues to do so until the ball ceases to fall because of some counter-acting force. The falling of the lead ball is a process which is the proximate $cause_1$ of the indenting of the bag since these two processes are coterminus. Similarly, the movement of the sculptor's hands is the proximate $cause_1$ of the changing in shape of the block of clay. In cases of proximate $causation_1$ it might seem arbitrary as to which event is designated as the cause and which as the effect. This, however, is not so; for only one of the two related events will have a prior cause (a proximate $cause_2$ or $_3$) which is independent of the other event, i.e. this prior cause could occur without this other event occurring. Thus the falling of the ball is said to be the cause since there is a prior cause for it, such as someone's releasing the ball or the ball being knocked off a shelf, which makes no reference to the indenting of the bag, but there is no prior cause for the indenting of the bag which makes no reference to the ball's falling on it. The movement of the locomotive is said to be the cause of the movement of the cars which are attached to it, since the movement of the locomotive alone can be accounted for in terms of prior causes, such as the manipulation of various levers in the locomotive or the stoking of its fire, etc. (We are ignoring the temporal lapse between the movement of the locomotive and the attached cars due to the elasticity of the connecting medium.)

We are concerned not only with the cause of the changing in shape of the sandbag and the block of clay but also with the proximate cause of the final or completed shape of the bag or clay. When we ask for the proximate cause of the final indentation of the bag (as opposed to the indenting of the bag, which is a process rather than a state), we want to know what event immediately preceded this state of affairs. We distinguish between a physical object or system changing because of some outside intervention and this object or system enduring in some manner, for example, having some enduring property or changing in some uniform manner. The being fully indented is an enduring state of the bag, which commences at the instant at which the outside intervention—the falling of the lead ball—ceases to operate on it. The falling of the lead ball is the proximate $cause_2$ of the fully indented state of the bag since it ends just when its effect begins. In Aristotelian terminology, we might say that the efficient cause consisting of the movement of the sculptor's hands ceases at the instant at which the form is embodied in the clay. Thus the movement of the sculptor's hands is the proximate $cause_1$ of the changing in shape of the clay and the proximate $cause_2$ of the completed shape of the clay.

Often when we inquire into the proximate cause, or, simply, *the* cause, of a puzzling or unusual event or state of affairs, we do not want to know either what happens simultaneously with this event or what is temporally contiguous with it, but rather what initiates some process which in the normal course of things terminates in this event. Here the proximate or immediate cause is temporally separated from its effect by the process that it initiates and which terminates in the beginning of the effect. The action of a man who poisons someone is the proximate $cause_3$ of his death, since this action initiates a process which *ceteris paribus* will terminate in the person's death. This requires that there not occur during the time in which this process goes on any abnormal event or voluntary intervention of a third party which, in the context under consideration, would be said to be the proximate cause of the person's death. For example, if A pushes B out of the 100th floor window of the Empire State Building A initiates a process which in the normal course of things will terminate in B's death. But if C should shoot B dead (killing him instantaneously) as he passes the 50th-floor window A's action is superseded as the cause of B's death because there has been a voluntary intervention of a third person which makes the situation non-standard. In law, C's subsequent action is said to negative A's action as being the immediate cause of B's death; A can be tried only for attempted murder.[1]

A *remote* cause of an event is a cause which, like a proximate $cause_3$, is temporally separated from its effect, but which, unlike a proximate $cause_3$, is the proximate cause of some sequence of causally related events one of which is, in the context under consideration, said to be the proximate cause (in one of the three senses given above) of the effect. While a proximate $cause_3$ initiates some process, or sequence of events, it is not the case that some phase of this process, or some member of this sequence of events, is, in the context under consideration, said to be a proximate cause of the effect. It is important to put in the phrase 'the

[1] This example appears in H. L. A. Hart and A. M. Honore, *Causation in the Law* (Oxford, 1959), p. 190. I am deeply indebted to this excellent book throughout this chapter.

context under consideration' to emphasize that we must not shift contexts, because then it could be the case that in the initial context some event is the proximate $cause_3$ of an event X but in some new context there is some subsequent event to this cause which is the proximate cause of X. For instance, in a legal context A's shooting B might be said to be the proximate $cause_3$ of B's death, but in a medical context the proximate cause would be said to be the deprivation of his blood cells of oxygen, which occurs subsequently to his being shot.

It is a gross oversimplification to speak of causality as a relation between two events, since this omits reference to the situation or state of affairs which must obtain if this causal connection is to hold. While common-sense causal generalizations do not specify in any detail what this situation is, we are nevertheless able to recognize when the situation is non-standard, and therefore the causal generalization does not apply. In a non-standard situation some necessary condition for the occurrence of the effect is absent or some condition sufficient for preventing the effect is present. Mill claimed that the set of all necessary conditions constituted the sufficient cause of an event, but this fails to take account of the common-sense distinction between a cause and a mere condition. This distinction is based on the fact that a cause, as distinguished from a necessary condition, is usually either an intentional action and/or an abnormal occurrence in relation to the subject-matter under inquiry. Oxygen is a condition but not a cause of a combustion, since oxygen is usually present; but if the usual state of affairs is one in which oxygen is absent, as in a vacuum tube, then the presence of oxygen in this tube could be said to be the cause of a combustion in this tube. In the course of this chapter, when it is said that one event is a sufficient cause of another, this will presuppose that there is a standard situation in which these two events occur, though we cannot specify in detail every necessary condition which makes up this situation.

The distinction between a cause and a mere condition explains why it is that we never speak of a remote cause as being sufficient. A remote cause is a proximate cause of some sequence of causally related event, the final one of which is the proximate cause of the effect. If we were to try and specify the standard situation in which a remote cause is sufficient, reference would have to be made to this sequence of causally related events, along with the situations in which each of them is a sufficient proximate cause of the event following it. Not only would our specification of this situation be fantastically complicated and extend over a long period of time but more important it would have to include reference to either an unusual event or an intentional action, viz., the final one in the sequence of causally related events. But by including an abnormal event or an intentional action as a condition that comprises a part of the standard situation we are violating our distinction between a cause and a condition.

The situation or state of affairs in which one event is a sufficient proximate cause of another will cover the period of time from the beginning of the cause to the end of the effect. The reason for this is that the presence or absence of some condition during this time could prevent the effect from occurring. This consideration helps to answer the Russell–Taylor paradox concerning the seeming impossibility of a cause ever being earlier than its effect.[1] For all they show is that the *situation* must extend to the end of the effect, not that the cause cannot precede its effect. The burning of logs in a fireplace is the proximate $cause_1$ of the simultaneous warming of the room only if certain conditions obtain during the occurrence of these two events, such as the room being sufficiently well insulated, no refrigerator being left open in the room, etc. The falling of the lead ball on a sandbag is the proximate $cause_2$ of the fully indented state of the bag only if certain things obtain *both* while the cause occurs (e.g. the ball is not so heavy that it will break the bag) and while the effect occurs or obtains (e.g. there is no counter-acting force, such as someone ramming his fist upward into the bag at the instant the ball ceases to fall). Similar considerations apply to cases of proximate $causation_3$, the only difference being that the cause is temporally separated from its effect.

II. NECESSARY CONDITIONS FOR A COUNTER-STIPULATION-EXAMPLE

Having said something about the distinction between a proximate and remote cause and between a cause and a condition, we are

[1] Russell, 'On the Notion of Cause', in *Mysticism and Logic* (London, 1950); and R. Taylor's two articles, 'Causation', *Monist*, 47 (1963), and 'Can a Cause Precede Its Effect?' *Monist*, 48 (1964).

now able to specify certain necessary conditions that must be fulfilled by any successful counter-stipulation-example to the analytic truth that a present event cannot have a past effect. In saying that these conditions are necessary it does not follow that satisfaction of them is sufficient for a successful counter-stipulation-example; in fact, it will be argued in Section III that any alleged counter-stipulation-example which fulfills these necessary conditions will fail because it will cause absurdities to break out in neighbouring concepts. The conditions which must be satisfied by any successful counter-stipulation-example of a present event *L* being a sufficient cause of a past event *E* are:

(1) The ordinary use of 'past' and 'future' must not be radically altered.
(2) It must be the case that every occurrence of an event of type *L* in a situation of type *S* has been preceded by an event of type *E*.
(3) *L* must be an intentional action.
(4) *E* must not be a proximate or remote cause of *L*.
(5) There must be no proximate cause of *E* that is earlier than *L*.

The meaning of, and necessity for, some of these conditions may not be readily apparent. Therefore, it is essential that each of these conditions be separately explained at some length, and its necessity justified.

(1) The necessity for this condition is that without it a counter-stipulation-example might fail through equivocation. When a person claims that some example justifies our saying that a present event causes a past event he must continue to use the words 'past' and 'future' in their familiar sense. The ordinary use of these words is based on certain logical and empirical criteria for distinguishing an asymmetry between the past and the future; an example of the former being that it is logically possible to remember the past and to anticipate the future, and an example of the latter being that our birth is in the past and our death in the future.[1] Any successful counter-stipulation-example requires a

[1] It might be thought that there are counter-examples to the claim that we cannot remember the future, such as 'I remember that there is a picnic tomorrow'. The latter is obviously elliptical for 'I remember that there is a picnic scheduled (planned) for tomorrow', in which case I am remembering my past state of knowledge and not a future event. The reason is that it could be true that I remember (in the success

background of normal causal sequences (e.g. in which a scar is later than the slashing, etc.) in which causes are not later than their effects, and in which it is the past that is remembered and the future that is anticipated. If we should decide to say that our birth is in the future and our death in the past, etc., and that we remember the future and anticipate the past, then we will have accomplished the trivial feat of merely interchanging the use of 'past' and 'future' in our language. This will require us to recast certain analytic truths, so that it now is an analytic truth that a present event cannot have a *future* effect. But then we would have to devise a new counter-stipulation-example to our old analytic truth which would consist of a present event causing a *future* effect. Backward causality necessarily must be the exception. One might object here that one of our logical criteria for distinguishing between the past and the future is that a present event can have future, but not past, effects, and to give this up is *ipso facto* to change our use of 'past' and 'future'. To meet this objection let us say that we must continue to use in our counter-stipulation-example a majority of both our logical and empirical criteria for distinguishing between past and future.

(2) The necessity for this condition is obvious. Part of the meaning of the claim that a concrete event *L* in a situation *S* causes another concrete event *E* is a generalization to the effect that if there is an occurrence of an event of type *L* in a situation of type *S* it will be accompanied by an event of type *E*. Thus data to the effect that numerous occurrences of events of type *L* in situations of type *S* are without exception accompanied by events of type *E* constitute evidence for the claim that a concrete event *L* in *S* causes a concrete event *E*. That a causal claim about two specific events entails a generalization about a class of events and a situation is perfectly consistent with the belief that there can be a causal connection between two unique events in a certain situation, that is, events that occur only once in the history of the

[1] For a more detailed discussion of this point see my article, 'Professor Ducasse on Determinism', *Philosophy and Phenomenological Research*, XXII (1961).

sense of 'remember') that there is a picnic tomorrow without there being a picnic tomorrow: the picnic which was scheduled or planned for tomorrow could be rescheduled. Similar considerations would hold for our memory of events occurring prior to our birth. For a discussion of these cases see Norman Malcolm's three lectures on memory in *Knowledge and Certainty* (Englewood-Cliffs, 1963).

universe.[1] If the counter-stipulation is to show that L in S is a *necessary* cause of the earlier E, then the data required would be that events of type E in situations of type S without exception have been followed by events of type L.

But the data required by condition (2) are neutral as to whether L is a sufficient cause of E or whether E is a necessary cause of L. In fact, it does not even establish that there is a causal connection between E and L: their conjunction on numerous occasions could be due to chance. This shows that causality cannot be analysed solely in terms of logical sufficiency/or necessity. To be a cause of is something more than to be a sign of. For this reason causal statements cannot be looked upon as Rylean inference tickets or as statements of functional correlation.

(3) In addition to events of type L in situations of type S never failing to be preceded by events of type E, it must be the case that L *brings about* E or *makes* E happen. This notion of *effectiveness* is an essential part of our ordinary concept of causality. The paradigm cases of causality are those in which an agent intentionally intervenes into some situation and brings about a change in it. The situation, for example, might involve a cart's being at rest, and then starting to move when an agent begins pushing it. These are paradigm cases of effectiveness—one event being made to happen or brought about by another. Because of the deeply entrenched nature of the analytic truth that a present event cannot have a past effect, we shall require that a counter-stipulation-example involve a paradigmatic case of causality; therefore, L must itself be an intentional action. If this strong condition is not fulfilled, there would be no compelling reason for saying that L in S is sufficient cause of E rather than that E in S is a necessary cause of L.

It is very difficult to find a case of a present event making something happen in the past because the concept of making involves that the process of making be temporally prior to its result, whether the result be a product or a state of affairs (or event). Just as a particular boat cannot be temporally prior to the process of making it, the state of exhaustion cannot be temporally prior to the process, such as running, that makes a person exhausted. Making some thing or making something happen involves the active control of some material or mechanism. Since ordinary verbs of making have built into them the posteriority of their results, we cannot say that we make something to have happened. However, there is available to us a weaker expression, 'bringing it about', which does not require the active control of some material or medium by an agent. Given a suitable counter-stipulation-example we might be justified in saying that someone brings it about that something has happened, even though he has not manipulated any levers, pulled any wires or chiseled any marble. Even with this weaker sense of causing something to happen we run into difficulty, since most ordinary action verbs, which involve bringing something about, require the posteriority of their effect. To find a counter-stipulation-example of an agent intentionally performing some action to bring it about that something has happened we shall have to either invent some new action verb or modify one that we have. We shall allow our alleged counter-stipulation-example to envisage any sort of bizarre counter-factual association of events and to invoke as many coincidental occurrences as required. While there is complete freedom to supply content to causal generalizations or laws there must not be too radical a departure from the form of causal generalizations or laws: that is, a successful counter-stipulation-example must retain most of the features of our present conception of a causal law or explanation.

It might be argued that in regard to any alleged counter-stipulation-example it can be shown that there is an incompatibility between condition (3) and (2), requiring that *L* never occur in *S* without being preceded by *E*. The argument is based on the introduction of experimental control in which the agent who can do *L* at will waits for some occasion on which *E* does not occur and then tries to do *L*. If he is successful condition (2) is violated, since there is an occurrence of *L* in *S* which is not preceded by *E*; and if he is unsuccessful after numerous attempts it shows, in violation of condition (3), that *L* is not an intentional action, that is, one that can be done at will. This argument is easily countered, for from the fact that on numerous occasions on which *E* did not occur the agent was unsuccessful in his effort to do *L* it does not follow that *L* is not an intentional action. On each occasion on which he tried and failed to do *L* there might have been lacking some necessary condition for his doing *L*, the absence of which was not a result of the non-occurrence of *E*. This may require us to introduce a number of unlikely coincidences to account for the agent's failure to do L on those occasions, but, as pointed out above, we allow our counter-stipulation-example to invoke as many coincidental occurrences as it requires. For example, E might consist in Jones dreaming that Smith will play *Petrouchka* on the piano the following day (call this event L). Smith is able to play the piano and also knows *Petrouchka* by heart. Smith waits for some occasion on which E does not occur and then attempts to do L on the following day. On one occasion his piano is stolen and on another thieves enter his home and tie him up just as he is about to strike the keyboard, etc., and none of these coincidences is caused by the non-occurrence of E. But from the fact that Smith did not do L on these occasions, it does not follow that he did not have the ability to do L on these occasions, for *ex hypothesis* he is able to play *Petrouchka* at will.

(4) This condition is necessary because it is an analytic truth that causality is asymmetric. Therefore, if E is a cause of L, L cannot be a cause of E. To give up this analytic truth along with the analytic truth that a present event cannot cause a past effect would depart too far from our ordinary concept of causality to count as a recognizable analogue. An example of an alleged counter-stipulation-example that violates this condition would be one in which an agent makes L happen because of his knowledge that E occurred. E might be a mentalist predicting that a man will ride into the night club in five minutes on top of an elephant. His stooge, who is planted outside the club on top of an elephant, hears E on a walkie-talkie radio, and five minutes later rides triumphantly into the club whistling the march from *Aida* (call this event L). Plainly E is a remote cause of L since the stooge does L be*cause* he has previously heard E. If we should persist in calling L the cause of E it would entail that E is God. The reason for this is that E is a cause of L and L in turn is a cause of E, in which case E is a cause of itself and therefore is God. One might quibble here that E still is not God since E is only a remote cause of itself.

It might be argued that even when collusion is absent, and therefore L is not caused by E, it can be shown that there is an incompatibility between condition (2) and (4). This argument is basically the same as the one presented in our discussion of condition (3), which is based on exercising experimental control over L so as to make it happen in the absence of E, thus refuting the claim that L is a sufficient cause of E.[1] Whereas the above argument claimed that repeated failures to do L in the absence of E showed that L is not an intentional action, this new form of the argument claims that it entails, contrary to (4), that E is a cause of L. The first premiss of this argument is that the data required by condition (2) is neutral as to whether L is a sufficient cause of E or whether E is a necessary cause of L. If E is a necessary cause of L then not-E is a sufficient cause of not-L. Therefore, if on occasions when E does not occur we are unable to make L occur, thus satisfying condition (2), it shows that not-E is the sufficient cause of not-L. And this violates condition (4) since it entails that E is a necessary cause of L.

The conclusion of this argument is a non-sequitur, for it is hardly not-E that is the sufficient cause of not-L. In fact it was not-E that was the cause of our trying to do L, and usually our effort to do something which we have the capacity to do is a positive, rather than a negative, causal factor in its happening. On each occasion when we fail to do L in the absence of E there could be a causal explanation, coincidental though it may be, for not-L which is independent of not-E. For example, we wait for some occasion on which the mentalist does not predict that a man will ride into the club in five minutes on an elephant and then attempt to ride into the club on an elephant. On one occasion we fail to get an elephant because the Hertz-Rent-an-Elephant people are all out of elephants, due to the Shriners being in town, and on another occasion the person riding the elephant falls off just as he is about to ride into the club, etc.

(5) Why is it necessary that there be no proximate cause of E which is earlier than L? The reason is that if there were a proximate cause of E which is earlier than L then there would be no causal determining of E left over for L in S to perform. In other words, if we already have a complete causal explanation of E in terms of an event(s) earlier than or simultaneous with E there is no need to invoke some later event L to causally account for E.

A rather ingenious argument might be advanced to show that condition (5) can never be satisfied. L is an intentional action for

[1] It might be argued that the introduction of experimental control of L violates condition (4), which requires that E not be a proximate or remote cause of L. This is not so, for if we were successful in making L happen when it is not preceded by E it would show only that not-E is on that occasion a remote cause of L (since it is our knowledge of not-E that causes us to bring about L), not that E is a cause of L.

which the cause might be certain psychological factors, such as the agent's desires, motives, etc.: and, granted the principle of determinism, there can be no limit placed upon how far back we can trace the causal ancestry of this cause of L. But this entails that we can always find some causal ancestor of L that is either simultaneous with or contiguous with E. But then there is a proximate cause of E which is earlier than L. The following diagram, using 't' to represent an interval (rather than an instant) of time and '⌒' to represent *is a proximate cause of*, illustrates this:

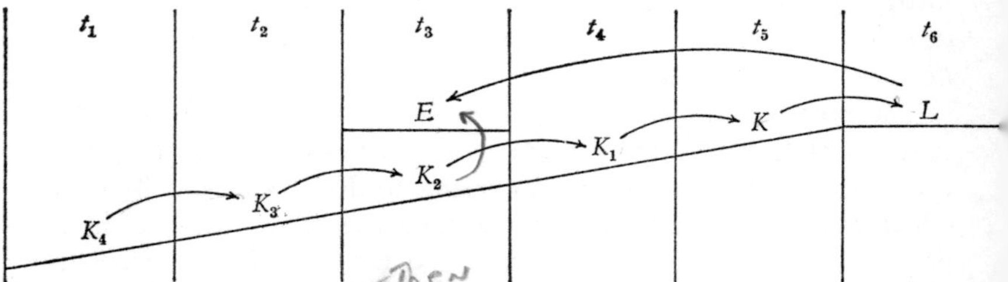

It is argued that in this diagram K_2 qualifies as a proximate cause$_1$ of E and K_3 qualifies as a proximate cause$_2$ of E, thereby failing to satisfy condition (5). The reason for this is based on the definition which we gave of a 'proximate cause' in section I. A 'proximate cause$_1$' is a cause (in some undefined sense) which is simultaneous with its effect, and a 'proximate cause$_2$' is a cause (in some undefined sense) which is temporally contiguous with its effect. Since K_2 is a cause of E and is simultaneous with E it is, by definition, a proximate cause$_1$ of E; and since K_3 is a cause of E and is temporally contiguous with E it is, by definition, a proximate cause$_2$ of K. It would be inconsistent with our definition of 'remote cause' to say that K_2, as well as K_3, are remote causes of E, for a 'remote cause' was defined as being both temporally separated from its effect *and* the proximate cause of some sequence of causally related events one of which is the proximate cause of the effect. While K_2 and K_3 satisfy the second conjunct of this definition, they fail to satisfy the first conjunct, requiring that a remote cause be temporally separated from its effect. One cannot counter this argument by changing the definition of 'proximate cause' so as to rule out the possibility of a proximate cause of an event being the proximate cause of some sequence of causally related events the final one of which is the proximate cause of this event. For this gives us a circular definition, e.g. A 'proximate cause_1' is a cause which is simultaneous with its effect and is not the *proximate cause* of some sequence of causally related events the final one of which is the *proximate cause* of this effect.

There is a way of meeting this argument. Let us assume that the concrete event K_2 occurred in a situation C. If the concrete event K_2 in C is a sufficient proximate cause of E this entails, not that there are repetitions of K_2-type events in C-type situations accompanied by E-type events (there can be a causal connection between unique events), but that whenever (if ever) events of type K_2 occur in a situation of type C they are accompanied by events of type E. All our counter-stipulation-example has to do in order to rule out the concrete event K_2 in C as a sufficient proximate cause of E is to stipulate that there is at least one occurrence of an event of type K_2 in a situation of type C that is *not* accompanied by an event of type E.[1] And similar considerations hold *mutatis mutandis* for K_3. To rule out in this manner any causal ancestor of L as being a proximate cause of E will require our counter-stipulation-example to make many stipulations concerning the way in which events occur, but we have agreed to allow complete freedom in the number and type of stipulations which are made concerning the *de facto* connections between events. Since there is no proximate cause of E earlier than L it follows that any remote cause of E which is earlier than L, such as K_4 in the above diagram, cannot be a proximate or remote cause of any event earlier than L which is the proximate cause of E. Rather, K_4 can be a remote cause of E only because it is a remote cause of L. And note that K_4 is not a remote cause of L in virtue of its being a remote cause of E since, in accordance with condition (4), E cannot be a remote or proximate cause of L. K_4 must be the proximate cause

[1] One might argue that since K_2 is a sufficient proximate cause of K_1, and K_1 a sufficient proximate cause of K, and K a sufficient proximate cause of L, therefore K_2 is a sufficient cause of L. And if this is so there could not be an instance in which an event of type K_2 occurs without being followed by an event of type L. And since L is a sufficient proximate cause of E it entails that E must occur every time K_2 occurs, in violation of the stipulation. This argument fails because it requires us to speak of a remote cause of L, namely K_2, as a sufficient cause of L. As pointed out in Section I, we cannot speak of a remote cause as sufficient because the concept of a situation in which it is sufficient has no application to a remote cause. K_2 is sufficient for K_1 in a situation C. And K_1 is sufficient for K in a situation D. But then it is possible for K_2 to occur in C and sufficiently cause K_1 but the situation in which K_1 occurs is not situation D, and therefore it is not sufficient for K.

of some sequence of causally related events which does not include *E* and one of which (in this case *K*) is the proximate cause of *L*.

Having done something to explain and justify condition (5), it can be shown why most of the counter-stipulation-examples that have been offered, and, in particular, all of those offered by Dummett in his two articles, have failed to satisfy this condition. Dummett, in his article on 'Bringing About the Past', gives an example of a tribe which has an initiation rite in which their young warriors go off on a six-day trip to prove their bravery. The young men are accompanied by elders of the tribe who act as witnesses. They spend the first two days journeying to the hunting area, on the third and fourth days the young warriors fight with wild animals and prove themselves brave or cowardly, and they travel homeward on the fifth and sixth days. The chief of the tribe dances during the entire six-day period to cause them to be brave. Now the brave conduct of the young men on the third and fourth days is the sort of event for which there are prior causes which together are a sufficient cause of this event, such as eugenics, education, wild animal combat training, promises of rewards and punishment, etc. Dummett is conveniently silent as to how these young men are raised: a tribe would be most remiss if it relied solely on the chief's six-day dance marathon to bring about the men's bravery and paid no attention to the way in which these men were reared. On any occasion when the men's brave conduct can be satisfactorily accounted for by prior causes there would be no need to invoke the chief's dancing on the fifth and sixth days as the cause, or even a part of the cause.

A similar objection can be made to another one of Dummett's examples in which:

> A man is observed regularly to wake up three minutes before his alarm-clock goes off. He often does not know when he goes to sleep whether or not the alarm has been wound, nor for what time it has been set. Whenever the alarm has been set and wound, but fails to go off because of some mechanical accident, which is later discovered, he always sleeps till very late.[1]

Now a person's waking up is the sort of event for which there are known prior causes, such as being shaken violently. Dummett admits as much when he says that on occasions when the alarm

[1] Dummett, 'Can an Effect Precede its Cause?' *op. cit.*, p. 32.

fails to go off the man over-sleeps. But he nevertheless wakes up without the alarm going off three minutes later: he is not doomed to a possibly endless sleep that can be ended only by the arrival of a princely alarm-clock that is about to go off in three minutes. On any occasion when there is a prior sufficient cause for the man waking up we would not invoke the subsequent ringing of the alarm-clock as the cause.

The objections which have been raised to Dummett's two examples could be met by stipulating that there are no proximate sufficient causes for the brave conduct of the men and the waking of the man which are earlier than, respectively, the chief's dancing on the fifth and sixth days and the alarm-clock ringing. This stipulation is in accordance with condition (5), which requires that there must be no proximate cause of *E* which is earlier than *L*. But condition (5) is ambiguous, admitting of a strong and a weak interpretation. The weak version, to be referred to as '(W5)', states: On every occasion in which an event of type *L* occurs in a situation of type *S* there must be no proximate cause of *E* which is earlier than *L*. The strong version, to be referred to as '(S5)', states: On every occasion in which an event of type *E* occurs there must be no proximate cause of it which is earlier than the subsequent occurrence of an event of type *L* in a situation of type *S*. The difference between these two versions of (5) is that (W5) permits an occurrence of an event of type *E* for which there is a proximate cause which is earlier than or simultaneous with *E*, provided only that an event of type *L* does not subsequently occur in a situation of type *S* on that occasion; whereas, (S5) rules out this possibility, since it requires that no occurrence of an event of type *E* can have a proximate cause earlier than the subsequent occurrence of an event of type *L* in *S*. It shall now be shown that (W5) involves a departure from the ordinary concept of a causal explanation and law, and it shall then be shown that (S5) is subject to the same difficulty, only not in as obvious a manner as (W5).

In Section I it was pointed out that the situation in which a cause *C* is sufficient for its effect *E* will cover the period of time from the beginning of the cause to the end of the effect. The reason for this is that the presence or absence of some factor during this period of time could negative *C* as the cause of *E* either by *preventing* *E* from happening or *superseding* *C* as the cause of *E* (e.g.

our earlier example in which the shooting of a man as he falls past the 50th-floor window supersedes his being pushed out of the 100th-floor window as the cause of his death). The description of the situation in which one event is a sufficient proximate cause of another must not refer to what does or does not obtain at any time *earlier* than the *beginning* of the cause or *later* than the *end* of the effect.[1] Obviously the temporal relation between a situation and two causally related events will be different when the cause is later than the effect, but we can derive the formula for the description of this situation by interchanging 'earlier' with 'later' and 'beginning' with 'end' in the preceding formula for the description of a situation in which a cause is not later than its effect. The description of the situation in which one event is a sufficient proximate cause of another (earlier event) must not refer to what does or does not obtain at any time *later* than the *end* of the cause or *earlier* than the *beginning* of the effect.[2] (W5) violates this formula because it requires that our description of the situation in which L is a sufficient proximate cause of the earlier E make reference to what does not obtain *earlier* than the *beginning* of the effect E, namely that there is no proximate cause of E which is earlier than E. If there is an earlier proximate cause of E it negatives L as the cause of E by superseding L. Thus the ringing of the alarm-clock is the sufficient proximate cause of the man's waking up three minutes earlier only if this earlier event is not preceded by his being shaken violently, etc. This is analogous to describing the situation of a normal prior proximate cause by reference to what does not occur *later* than the *end* of the effect: shaking a man violently is the cause of his waking up only if in the situation there is not a ringing of his alarm-clock three minutes after he wakes up.

It might be thought that this difficulty can be avoided by replacing (W5) by (S5). Since (S5) stipulates that there must be no proximate cause of an event of type E other than L, and thus none which is earlier than E, we need not describe the situation in which L is a sufficient cause of E by reference to the non-occurrence of certain other possible proximate causes of E earlier

[1] This holds true even if a proximate cause$_3$ is not linked with its effect by some process or sequence of events: in Section I such causality at a temporal distance was not countenanced.

[2] This formula is neutral as to whether backward causation$_3$ operates at a temporal distance. In most counter-stipulation-examples it does operate at a temporal distance.

than *E*. If (S5) is satisfied then there is *as a matter of fact* no proximate cause of *E* earlier than *E*, and therefore there is no event earlier than *E* whose occurrence must be ruled out in the description of the situation in which *L* is causally sufficient for *E*.

This superiority of (S5) over (W5) is, however, illusory. The question is not whether there is *in fact* any earlier proximate cause of *E* but what we would say if (as is logically possible) there were one. To understand the meaning of an expression we must elicit its rules of use, which permit us to use the expression on certain sorts of occasions and in certain sentential contexts, *but not in others*. The rules of use specify a set of permissible as well as prohibited moves. In eliciting the meaning that the word 'cause' has in a counter-stipulation-example which satisfies (S5) we must discover its rules of use, and we discover these not only by noting on what sort of occasions it is used but also by seeing on what sort of occasions it would not be used. Thus, the question, 'Would you call the concrete event *L* the cause of *E* if on that occasion there were a prior proximate cause of *E*?' is relevant to determining the meaning that 'cause' has in a counter-stipulation-example satisfying (S5). That such a state of affairs does not in fact occur, which is what (S5) stipulates, is irrelevant to how the word 'cause' is being used. The fact that (S5) requires us to answer that *L* is the cause of *E* only if on that occasion there is no proximate cause of *E* which is earlier than *E* shows that 'cause' is being used in the same way as it is in an example that satisfies (W5). And therefore the same difficulty which applies to the use of 'cause' in a counter-stipulation-example satisfying (W5) holds for one satisfying (S5).

A defender of a counter-stipulation-example satisfying condition (W5) or (S5) might try to minimize the extent to which his use of 'cause' departs from its ordinary use. He might argue that the reason why we do not have to characterize the situation in which an event is causally sufficient for a non-earlier effect by reference to what does not happen later than the end of the effect is due only to the contingent fact that we do not yet have any causal laws involving backward causation. But if such laws were to be established then if a given event could have both a prior and a posterior cause, we would have to specify the situation in which some earlier event causes it by reference to the non-occurrence of its posterior cause on that occasion. The difficulty with this reply is that it begs the question, for it attempts to eliminate the difficulty attendant upon a counter-stipulation-example by presupposing that we already have, or could have, backward causal connections, which is the very thing to be proven by a counter-stipulation-example. I do not know how serious a difficulty this is for a counter-stipulation-example, for it is not clear whether the queer way in which the situation is described in such an example is so radical a departure from ordinary usage as to disqualify *L* as being the cause of *E*. We shall not linger on this difficulty, for we now hope to show that there are far graver conceptual difficulties connected with any counter-stipulation-example satisfying conditions (1)–(5).

III. ABSURDITIES RESULTING FROM A COUNTER-STIPULATION-EXAMPLE

We shall now devise a counter-stipulation-example satisfying conditions (1)–(5) for the purpose of showing that it involves a radical departure from the ordinary concept of causality. It will be shown that such an example causes absurdities to break out in certain neighbouring concepts with which the concept of causality has logical liaisons, such as action, intention, deliberation, reason, memory, experimentation, responsibility and punishment. These absurdities can be resolved only by radically altering these concepts. And since the meaning of 'cause' depends upon its logical relationships with neighbouring words in the language, the changes which we are forced to make in the meaning of the latter are *ipso facto* changes in the meaning of the former.

Let us devise a counter-stipulation-example, which shall be called the *Case of the Retro-Warning*. On numerous occasions Jones learns that his friend Smith was in a dangerous situation at some earlier time. He learns of this in perfectly ordinary ways—records, testimony of observers, collation of disparate memories, etc.—but he does not know what was the eventuation of this dangerous situation nor whether Smith was warned of the danger by anyone. On each of these occasions Jones utters a 'retro-warning' (call this event *L*); for example, he says, 'Smith, this is Jones talking to you from your future. There's a time-bomb in the desk drawer!' And on each of these occasions it is the case that Smith has had a para-normal auditory experience in which he hears the exact words that Jones subsequently utters said in Jones's voice (call this event E). The present event L causes the past event E because it is reasonable, so it is argued, to say that Jones does L in order to bring about that E has happened. This example satisfies all of the five necessary conditions. (1) The use of 'past' and 'future' is not radically altered, for it still is the past that can be remembered, etc. (2) L never occurs without being preceded by E. (3) L is an intentional action. (4) E is not a cause of L. (5) E is called 'paranormal' because it has no prior proximate cause, and, in fact, no proximate cause prior to L, thus satisfying (S5).

It is apparent that the Case of the Retro-Warning requires that numerous conceptual changes be made. Since Jones can perform an intentional action so as to bring about a past event it is necessary, for the sake of consistency, to alter our present concept of intention so that it is now possible for an agent to have intentions in regard to the past as well as the future. Moreover, we must also allow for *intentional knowledge* of the past as well as the future. An agent has intentional knowledge not solely on the basis of his intention to perform some action but also in virtue of possessing certain non-intentional knowledge. More precisely:

(i) An agent has intentional knowledge of one of his actions X if, and only if, he has an intention to do X and has non-intentional knowledge that he can do X.

(i′) An agent has intentional knowledge of an event Y which is not one of his actions if, and only if, he has intentional knowledge of one of his own actions X and knows that X is sufficient to bring about Y.

In the Case of the Retro-Warning Jones can have intentional knowledge not only of his action L but also of E, since he knows that L is sufficient to bring about E.

An agent gives up an intentional knowledge claim under one of two conditions: (*a*) he finds out either that he cannot carry out his intention to do X or that X is not sufficient to bring about Y; or, more commonly, (*b*) he finds good reasons for not doing X or for not bringing about Y. The 'good reasons' have to do with the desirability of X (Y) and its consequences as contrasted with the desirability of other possible courses of action. A peculiarity of intentional knowledge, due to the fact that an agent ceases to have intentional knowledge when he gives up his intention, be it under condition (*a*) or (*b*), is that that an agent has intentional knowledge that *X* will occur does not entail that *X* will occur. Some would question the propriety of speaking of intentional 'knowledge' as knowledge, while others might say that this is a counter-example to the claim that that *X* knows that *p* entails that *p*.

To allow for the possibility of backward intention it is necessary that our concept of deliberation be altered so that it is possible to deliberate about the past as well as the future. The necessity for this conceptual reform is due to the analytic truth that an intentional action is one which can, although it need not, be preceded by a period of deliberation about whether to perform this action.

At this point it might be asked, 'Granted that we must permit backward intention and deliberation so as to accommodate backward causation, but what is so conceptually absurd about backward intention and deliberation?' What I shall try to show is that this change in our concept of intention and deliberation *radically* alters our ordinary concept of the past, which is in violation of condition (1) for a successful counter-stipulation-example. Before doing this it is necessary to articulate a crucial analytic truth about our concept of the past. It is:

(ii) An event is past if, and only if, it is logically possible for any person who is neither in doubt about the occurrence of this event nor believes that it did not occur to have a trace of it.[1]

For a person to have a trace of an event it is required that the person both be in possession of a trace of this event and know on the basis of this trace that this event occurred: a person does not have a trace of an event if he merely has, unknown to himself, a photograph of this event stuck in his back pocket. On the basis of a trace a person can make a non-inferential knowledge claim about what happened. It is because of the intentional nature of the concept of *having a trace of* that it is necessary to restrict the conditions under which it is logically possible for a person to have a trace: it is not logically possible for a person to have a trace of an event and also either be in doubt about its occurrence or believe that it did not occur.

I have stated analytic truth (ii) about our concept of the past in terms of the generic concept of trace, rather than the more

[1] We are counting non-occurrences and negative states of affairs as events.

determinate concept of memory, since it could plausibly be maintained, Plato and Bridie Murphy not withstanding, that it is not logically possible for a person to have a memory of an event occurring prior to his birth, although he could have a trace (such as a photograph) of this event.

There is an important analytic truth about our concept of having a trace of, which could not be given up without *radically* altering our concept of trace and as a consequence our concept of the past. It is:

(iii) The only logically relevant reason for challenging or giving up a knowledge claim based on an ostensible trace is evidence which indicates that the ostensible trace is unveridical.

If an agent claims to have a trace of *X* we could legitimately challenge his claim by producing evidence that *X* did not in fact occur. A person gives up or begins to doubt a claim based on his ostensible memory when he finds new evidence which indicates that his ostensible memory is unveridical.

I shall now try to show that permitting deliberation and trace to work the same side of the street requires us to give up analytic truth (iii), thereby radically altering our concept of the past and rendering the Case of the Retro-Warning unacceptable as a counter-stipulation-example. Let us suppose, as is logically possible, that Jones along with several other persons were eyewitnesses on some occasion to the non-occurrence of *E* (*E* being Smith's paranormal auditory experience of being retro-warned): they were standing in the room with him and saw the bomb detonate in his face. Subsequently they remember the non-occurrence of *E* and also have other traces of this negative event, such as motion pictures and a tape-recording. Jones certainly is in a position to say that he knows that *E* did not occur; for he has more than ample traces to back up his knowledge claim. However, all of a sudden Jones begins to deliberate about whether to do *L* so as to bring about *E* in the past. At the moment Jones begins to deliberate he is logically required to give up his knowledge claim based on his ostensible memory and to call into question the knowledge claims of others based on ostensible traces of not-*E*. The reason for this is due to the following analytic truth about the concept of deliberation:

(iv) It is not logically possible that an agent deliberate about whether to do X and at the same time either know that X will occur or know that X will not occur.

A corollary to (iv) is:

(iva) It is not logically possible that an agent deliberate about whether to do X and at the same time know of the non-occurrence of some event of which X is a sufficient cause.[1]

It would be absurd for Jones to say that he is now deliberating about whether to do L but already knows that L will (will not) occur or that he is now deliberating about whether to do L so as to bring about E but already knows that E does not occur. If Jones already has such knowledge, he is engaging only in a mock deliberation.

What has happened is that Jones has had to give up his knowledge claim based on ostensible traces of not-E, not because he has discovered any new evidence to indicate that E did occur and that therefore these ostensible traces are unveridical, but merely because he has begun to deliberate about doing L so as to bring it about that E did occur. This is in violation of analytic truth (iii) that the only logically relevant reason for challenging or giving up a knowledge claim based on an ostensible trace is evidence which indicates that the ostensible trace is unveridical. To see exactly why this is in violation of (iii) it is necessary to make explicit the following analytic truth about the concept of deliberation:

(v) An agent's deliberation about whether to do X so as to bring about Y is concerned with reasons for doing X so as to bring about Y and not with evidence that Y in fact will occur.[2]

[1] It is logically possible for an agent to deliberate about whether to do X and at the same time know of the occurrence of some event of which X is a sufficient cause since there could be a sufficient cause of this event other than X. Also, someone beside himself might be able to do X so as to bring about this event. For example, Jones may be part of a conspiracy, each party to which has vowed to blow up the airplane on which they all are passengers. It is possible for Jones to deliberate about whether he should detonate his bomb so as to blow up the airplane even though he knows that the plane will blow up: what he is deliberating about is whether he should be the person to blow it up.

[2] Evidence to the effect that he can do L and that L is sufficient to bring about E can figure in the agent's deliberation, but this evidence by itself does not enable him to infer the occurrence of either L or E.

To deliberate about whether to do X is not to try and discover evidence that X will in fact occur; rather it is concerned with reasons for doing X, based on the desirability of the consequences of X over other possible courses of action. It would be absurd for someone who has just been deliberating about whether to do L so as to bring about E to say that he has made up his mind to do L because he has found conclusive evidence to the effect that E in fact will occur. Deliberating is not trying to infer what one will do. 'What ought I to do?' is not a request for a prediction of what I shall do but for good reasons for doing one thing rather than another. Therefore, Jones's giving up his knowledge claim (based on ostensible traces of not-E) in the face of his beginning to deliberate about the desirability of bringing E about is due to his consideration of *reasons* and not *evidence* for these ostensible traces of not-E to be unveridical. Examples of such reasons would be: Smith is not such a bad chap; his death would leave his wife and seven children destitute; we need Smith to play catcher in the company softball game next week, etc. Giving up analytic truth (iii) leads to a paranoic concept of the past, since an agent legitimately can challenge knowledge claims based on ostensible traces merely because he begins to consider good reasons for these claims to be false. There could hardly be a more radical change in our ordinary concept of the past.

There are many moves which a defender of backward causation might make at this point. First, he might stipulate, as he has a perfect right to, that on every occasion on which Jones deliberates about making E to have happened in spite of there being numerous ostensible traces of not-E it is subsequently discovered that all of these traces are unveridical; new traces are unearthed which show that all the eye-witnesses were having delusions and that the motion pictures of not-E are phonies, etc. This, however, is irrelevant, since it still remains the case that Jones is required, in violation of (iii), to give up a knowledge claim based on ostensible traces because of reasons rather than evidence for its being false. It is also unavailing to stipulate that it does not in fact ever happen that Jones begins to deliberate about making E happen when he and others have ostensible traces of not-E. The kind of absurdity that has been shown to follow from the assumption of backward causation is a *conceptual* rather than an *empirical* one. As pointed out on page 144, questions concerning what a defender of backward causation would be required to say if certain logically possible things were to happen are relevant to determining what he means by 'cause', 'deliberate', 'trace', etc. The above refutation of the Case of the Retro-Warning is based on a description of a logically possible state of affairs, and in eliciting the queer way in which Jones would have to speak in such a state of affairs it reveals that his concept of trace and therefore of the past is *radically* different from the ordinary one.

A defender of backward causation might question the supposition that it is logically possible for Jones to begin to deliberate about making *E* to have happened when he and others have ostensible traces of not-*E*. If such a supposition is not logically possible then Jones is spared the absurdity of having to give up his non-intentional knowledge claim based on these ostensible traces when he begins to deliberate about doing *L* so as to make these ostensible traces unveridical. The reply is that if we want, as would a defender of backward causation, backward deliberation and intention to be a true analogue to ordinary future deliberation and intention then it should be logically possible for an agent to give up a non-intentional knowledge claim because he begins to deliberate. An agent can claim to have non-intentional knowledge that *M* will not occur and then withdraw this claim, as he would be logically required to, when he begins to deliberate about doing *L* so as to bring about *M*. By doing *L* he does not change the future —what will be will be—but he makes it different than it would be if he were not to do *L*. Analogously, it should be possible for Jones to make a non-intentional knowledge claim that *E* did not occur and then withdraw this knowledge claim when he begins to deliberate about doing *L* so as to make *E* to have happened. What has been has been, but by doing *L* Jones makes the past different than it would have been if he were not to do *L*.

It might be claimed that it is only a contingent fact that there is no future analogue to trace. If, as is logically possible, we were to have a faculty which is a future analogue to memory, such as precognition, then we could have non-inferential knowledge of the future as well as the past.[1] In such a case an analogous logical clash would arise between ordinary forward deliberation and

[1] I am using 'knowledge' here in such a way that that *X* know that *p* entails that *p*. Intentional knowledge, as pointed out above, does not qualify as knowledge in this sense.

precognition as that which was shown to arise between backward deliberation and trace: there could be a situation in which an agent calls into question the veridicality of one of his ostensible precognitions of the future merely because he begins to deliberate and not because he has come upon any fresh evidence to indicate that the future is different than are the ostensible precognitions of it.

The obvious retort is that since the existence of precognition would result in a logical absurdity it is not a *contingent* fact that we do not possess such a faculty. Relative to our present conceptual system it is logically impossible that there be such a future analogue to memory. If we are to allow for the possibility of precognition it is necessary to jettison a family of concepts having to do with agency, such as deliberation and intention. It is not logically possible for an agent to precognize without interruption the decisions he will make as a result of future deliberations, whereas it is logically possible for an agent to remember without interruption the decisions he made as a result of past deliberations.[1] If an agent were to precognize without interruption the decision he will make as a result of a future deliberation he would know what decision he will make when he deliberates in the future, and this violates analytic truth (iv) that it is not logically possible that an agent deliberate about whether to do X and at the same time either know that X will occur or know that X will not occur. This brings out a crucial disanalogy between memory and precognition. It also shows that determinism, interpreted as the doctrine that an *agent* in principle could retrodict and predict every event on the basis of instantial premisses and scientific laws, is false.

In addition to radically transforming our concept of the past, the Case of the Retro-Warning also requires a drastic change to be made in our concept of experimentation. According to this Case, event L is the sufficient cause of a past event E. A causal claim

[1] An agent remembers without interruption an event X if between X and his present memory there is no lapse in his state of remembering X, as there would be if he were to suffer from amnesia during this time. Analogously, an agent precognizes without interruption an event X if between X and his present precognition there is no lapse in his state of precognizing X, as there would be if he were to suffer from some future analogue to amnesia—call it 'aforeknowlia'—during this time. The reason for requiring uninterrupted precognition is that otherwise an agent could now precognize the future occurrence of M but subsequently suffer from aforeknowlia and then honestly begin to deliberate about whether to do M.

should be amenable to experimental verification, although it may not be practically feasible to do so. The causal claim that an event *C* in a situation *S* causes event *E* is experimentally verified by making *C* happen in *S* and then observing what results. An agent's claim to be able to bring about *M* by doing *L* is experimentally verified by having him do *L* under controlled conditions and then waiting to see—to directly verify—whether *M* happens. Jones's causal claim to be able to bring about a *past* event *E* by doing *L* cannot be verified in this manner, for while we can wait for him to do *L* we cannot wait to directly observe whether *E* happens. In this case what we must 'wait for' after he does *L* is not the occurrence of *E* but indirect evidence of *E*'s having occurred in the past. This indirect evidence consists of traces. Traces, by definition, are enduring states of an object: the footprints in the sand, the marks on the photographic plate, as well as a person's memories, are enduring states of an object. Were traces not enduring states there would be no basis for an agreement in judgment between inquirers existing at different times subsequent to the event about which judgment is made. This means that all of the traces of *E* (not-*E*) that are in principle available to us after Jones does *L* are also in principle available to us before Jones does *L*. Thus we need not wait for the experiment to be performed in order to verify its outcome: before Jones does *L* we are in just as good a position to verify whether *E* occurred as we are after he does *L*. Jones is performing only a mock experiment.

Jones might say that the ostensible traces of *E* (not-*E*) cannot be trusted until after he does *L*. But if we cannot trust ostensible traces before he does *L* why should they be any more trustworthy after he does *L*? He might say that once he forms an intention to do *L* he has intentional knowledge of *E* and therefore has a basis on which to determine the veridicality of these ostensible traces. This is viciously circular since, according to analytic truth (i′), Jones has intentional knowledge of an event *E* which is not one of his actions if, and only if, he has intentional knowledge of one of his own actions *L and knows that L is sufficient to bring about E*. To know that *L* is a sufficient cause of *E*, Jones would need evidence which is independent of his intentions. If Jones should use his intention to do *L* as the decisive factor in determining whether *E* occurred he would sever himself from the rest of the scientific community. He would have a private science. He could go on cherishing the belief that he brings about E by doing L, only he probably would do so in a padded cell.

Backward causation also creates havoc for our concepts of responsibility and punishment. Let us suppose that E is a highly undesirable event and there is a law against bringing it about, anyone who does so intentionally being culpable and subject to punishment. On some occasion E happens, and, in accordance with the above Case of the Retro-Warning, it has no proximate cause other than Jones's subsequent action of doing L. Therefore, since an illegal event has occurred which is a result of Jones's intentional action L, we can hold him responsible—culpable—for this and punish him accordingly. Thus we are punishing him for some action that he has not yet performed. But part of the justification of punishment is to reform the character of the criminal so that he will not perform illegal acts in the future. However, this justification cannot apply to our punishment of Jones, for if the punishment does succeed in reforming Jones so that he will not do L in the future, then we will have punished him for a crime that he does not commit.

VIII
FUTURE CONTINGENTS

THE preceding chapter took a synoptic view of a family of interdependent logical asymmetries between the past and the future concerning causality, action, intention, deliberation, trace, knowledge and responsibility. In this and the following chapter these asymmetries will be appealed to in arguing for tenet (3) of the *A*-Theory, which holds that the past and future differ ontologically, the future being open and the past closed. In this chapter it will be argued that one sense in which the future is open is that it contains genuine possibilities or contingents.

In what sense of 'possibility', if any, is it true that there are future but not past possibilities? There are at least three senses of 'possible' in which there is no logical asymmetry between the past and future. Obviously *logical* possibility contains no asymmetry between past and future times, the reason being that it contains no reference at all to time. A second sense of 'possible' which, although it does contain a reference to time, nevertheless contains no asymmetry between past and future is the *epistemic* sense of 'possible'. An epistemic possibility involves a reference to our state of knowledge. 'It is possible that *S* was (is, will be) ϕ' means 'The evidence now available does not indicate that *S* was not (is not, will not be) ϕ'. What is epistemically possible at one time may not be epistemically possible at another time since the available evidence can vary from time to time. Another sense of 'possible', also containing no asymmetry between past and future, is the *pure capacity* sense. This involves reference to the capacity or ability that some person or thing possesses but does not specify or entail that the conditions necessary for exercising this capacity obtain. It might be said that it was possible for Dempsey to have beaten Louis, which means that given the capacities and abilities of these two fighters, had they fought (while each was at his peak) Dempsey would have won. Notice that this pure capacity sense of 'possibility' does not entail a corresponding epistemic possibility, for it is not possible that Dempsey defeated Louis, since it is known that they never fought.

There is a sense of 'possibility', to be called 'possibility for action or reaction', which is asymmetric between past and future. A possibility for action arises in a situation in which an agent can choose between, and therefore deliberate about, two or more alternative courses of action: its usual linguistic formulation is, 'It is now possible for *A* to do *X*', in which '*A*' refers to an agent and '*X*' to an action. This can be analysed roughly as, '*A* now has certain abilities and capacities and conditions are now such that if *A* chooses to do *X* then *X* will occur.' The predicate infinitive 'to do *X*' is used rather than an expression referring to an event, such as '*X*ing' or 'doing *X*', because a non-committal reference is made to the future: 'to *X*' or 'to do *X*' does not refer to an event since we cannot ask, 'When, or during what period of time, does to do *X* occur?' When '*A*' does not refer to a person, no reference is made to *A*'s choice. This is a possibility for reaction rather than action, the analysis of 'It is now possible for the dike to crumble' being 'The dike is now in a certain state or condition such that if certain things were to happen to it then it would crumble'. Some statements in which '*A*' refers to a person also express a possibility for reaction; for instance, 'It is now possible for the patient to die', assuming that we are not referring to the possibility of his committing suicide, means that the patient is now in such a condition that if certain other things were to happen he would die.

Possibilities for action and reaction not only can vary from time to time (since a thing's capacities and abilities, as well as the opportunities for the exercise of them, can vary with time) but also are asymmetric between past and future, these asymmetries resting on the asymmetries of the preceding chapter. Since a possibility for reaction makes reference to what effects would follow if certain things were to happen now, there can be future but not past possibilities for reaction, it being an analytic truth that a present event cannot have a past effect. And since a possibility for action refers to what would happen if an agent were to choose a certain course of action, such a possibility must refer to the future, it being an analytic truth that a person can deliberate, intend and choose only in respect to the future. A future contingent, as the term has often been used in the history of philosophy, is an event whose occurrence depends on how some agent will choose. It is only because agents have possibilities for action that there are future contingents. The question that will now be taken up is whether the existence of future contingents is compatible with truth being timeless. This is often referred to as the problem of the sea battle tomorrow.

Let us sharpen up the problem by sketching two diametrically opposed theories of truth and contingency. The first theory, to be called the 'Temporalistic Theory of Truth', is held by many *A*-Theorists. Aristotle has been interpreted by the Stoics, Epicureans, as well as by Lukasiewicz and Prior in modern times, as holding this view. It runs as follows:

> A prediction about a future contingent, such as a sea battle tomorrow, is neither true nor false now, even although the logical disjunction formed from this prediction and its contradictory is true *as a whole*: the disjunction, 'Either there will be a sea battle tomorrow or there will not be a sea battle tomorrow', is true as a whole, even although neither one of its disjuncts is true now. If one of the two disjuncts was definitely true (or false) now it would entail fatalism—would entail that nothing we do can make any difference in regard to the sea battle happening or not happening.

Aristotle, according to some interpretations, felt compelled to deny the universal application of the law of the excluded middle, which requires that every proposition be true or false, for a proposition can be said to be 'true', 'false', or 'neither true nor false'; and moreover a proposition, such as a prediction about a future contingent, can change in its truth-value from being neither true nor false to being true (false).[1]

The opposing theory, to be referred to as the 'Timeless Theory of Truth', is held by most *B*-Theorists. It runs as follows:

[1] For an excellent discussion of the various interpretations of Chapter 9 of Aristotle's *de Interpretatione* see Nicholas Resecher's 'Truth and Necessity in Temporal Perspective', in *The Philosophy of Time*, edited by Gale.

> Truth is a timeless property of statements or propositions. Since a statement's assumption of its truth-value does not admit of temporal distinctions it is nonsensical to speak of a statement as 'true now' or as 'becoming true', as does the Temporalistic Theory of Truth. This Theory also errs in claiming that a disjunction, such as '*p* or not-*p*', could be true as a whole even although neither one of its disjuncts is true now. In addition to the absurdity of predicating truth in a tensed manner, this claim is subject to the folly of thinking that the truth-functional statement '*p* or not-*p*' could be true even though neither of its disjuncts is true. The claim that there is an incompatibility between the existence of future contingents and the timelessness of truth rests on confounding logical determinism with fatalism. Logical determinism is the harmless tautology that what will be will be, but this does not entail the fatalistic doctrine that what will be will be of necessity, in the sense that that our choices and efforts make no difference. Because it is either true or false that we shall perform certain actions in the future does not entail that we are forced or compelled to do these things—that it is not now both within our power to do these things and also within our power to desist from doing them.

The fact that the controversy between the Temporalistic and Timeless Theories of Truth has raged for over two thousand years, with able philosophers lining up on different sides, indicates that each Theory probably has important insights into the logic of 'true'. I shall try to show that these opposing Theories can be reconciled if their claims are properly reformulated. My strategy will be to argue that the Temporalistic Theory is basically right if its claims are reformulated so as to apply to the pragmatic or contextual implications of 'true', while the Timeless Theory is basically right if its claims are restricted to the semantic meaning of 'true'. What will emerge is that expressions such as 'true now', 'neither true nor false now' and 'become true' have a use in ordinary language because of the pragmatic implications of 'true'.

Is truth a timeless property of statements, as contended by the Timeless Theory of Truth, or is it a property which a statement acquires at some time and retains for ever after, as the Temporalistic Theory of Truth holds? I believe that the Timeless Theory is basically right, provided it applies only to the semantic meaning of 'true', but that its manner of justifying this claim is wrong:

it is just plain false that 'true' is never predicated in a tensed way or in a way that makes reference to a specific time. For our purposes there are two important uses of tensed predications of 'true'. (*a*) There is a metalinguistic use of 'is now (was, will be) true' in which an *A*-statement is said to be true: such a metalinguistic statement will be referred to as an 'endorsing statement', but in calling it such I do not want to commit myself to any of the things that Strawson says in his mooted performatory theory of truth.[1] (*b*) A second use for tensed predications of 'true' is in a metametalinguistic statement about an endorsing statement to the effect that relative to available evidence we are warranted (not warranted) in making this endorsing statement. What I hope to show is that this metametalinguistic use of 'true' has the function of making reference to what is pragmatically implied by endorsing statements. First, I shall try to resolve the tension between the tensed predication of 'true' in endorsing statements and the timelessness of truth.

(*a*) Our endorsements of *A*-statements are often themselves tensed. The 'is true' in the endorsing statement, 'It is true that *S* is ϕ', is in the present tense, since it is meaningful to substitute 'was' or 'will be' for it. The reason for tensing such an endorsement is to indicate whether the *A*-sentence occurring in the noun 'that'-clause can now be used to make a true statement or could be so used in the past or the future. For example, to say 'It will be (was) true that *S* is ϕ' is to indicate that the *A*-sentence '*S* is ϕ' can be used in the future (past) so as to make a true statement. The occasion for making such a double perspective endorsement is when we envisage the earlier or later occurrence of some actual or possible speech-act involving the use of a sentence synonymous with that one in the 'that'-clause. An endorsing statement, whatever else it may do, at least makes or entails a statement that is

[1] See P. F. Strawson, 'Truth', *Analysis*, 9 (1948–49). The metalinguistic use of 'true' may have the illocutionary force of an endorsement, but this is compatible with its also having a descriptive or constative function, such as that of describing a relation of correspondence between a statement and an event or state of affairs. 'I believe', for example, in many uses has both the illocutionary force of a humble operator that indicates that our statement is not to be taken as something for which we can give much evidence, if any, and also the descriptive function of describing our state of mind. What shall be said about 'true' will be neutral between the correspondence and coherence theories of truth, though it will rule out certain pragmatic accounts.

logically equivalent to the one being endorsed.[1] If I want to now make a statement logically equivalent to the one made by someone who in the past (future) uses a sentence synonymous with 'S is ϕ' I would now have to use a sentence differing in tense from this A-sentence, namely 'S was (will be) ϕ'. My endorsing statement, 'It was (will be) true that S is ϕ', does entail that S was (will be) ϕ; and therefore my endorsement does make or entail a statement logically equivalent to the one that is being endorsed.[2]

In some tensed endorsing statements it is necessary, for purely grammatical reasons, to employ in the noun 'that'-clause a sentence that is not synonymous with the one used in making the statement being endorsed. For example, in endorsing a prediction made yesterday through the use of the sentence 'S will be ϕ tomorrow' I would have to say 'It was true yesterday that S would be ϕ today'. First, there is the shift to the subjunctive in the 'that'-clause: such a shift is not required when we endorse the statement made by a future use of a past tense sentence, e.g. 'It will be true tomorrow that S was ϕ today'. Second, there is the shift to the temporal indexical sign 'today', which expresses the temporal perspective of the endorser. Were we to say, 'It was true yesterday that S would be ϕ tomorrow', it would entail that S will be ϕ tomorrow, rather than that S is ϕ today.

It is always possible to paraphrase a tensed endorsing statement so as to eliminate a tensed predication of 'true', but it requires that some other A-expression be used. For example, instead of saying 'It was true that S is ϕ' we could say 'The statement made by a past use of a sentence synonymous with "S is ϕ" *is* true', which uses the A-expression 'past'. The sentences used in making both of these endorsing statements are subject to exactly the same temporal restrictions in their use for making a true statement: an endorsing statement is itself true or false. A similar distinction should be made between A-endorsing statements and tensed endorsing statements as was made between A-statements and tensed statements. Just as it was important to realize that there are numerous grammatical techniques other than the use of

[1] The reason why it is not required that it make or entail the *same* statement as the one being endorsed will emerge in Chapter X.

[2] My treatment is based on Wilfrid Sellars' excellent discussion in his 'Time and the World Order', *op. cit.*

tensed verbs for making an *A*-statement it is important here to realize that there are numerous grammatical devices for making an *A*-endorsing statement.

It should be noted that statements endorsing a non-*A*-statement are not themselves *A*-endorsing statements. The reason for this is that since a non-*A*-statement is made through the use of a sentence which cannot be used at two different times so as to make statements differing in truth-value the question of when this sentence can be used to make a true statement does not arise. Thus when we endorse the statement that two and two *is* four there is no need for us to suitably tense or employ some other *A*-expression in our endorsing statement; in fact, it would be nonsensical to say 'It is now (was, will be) true that two and two *is* four'.

The fact that statements endorsing an *A*-statement use 'true' in a tensed manner is compatible with the timelessness of truth. That truth is timeless is best brought out by considering what is entailed by an *A*-endorsing statement. Let 's_1', 's_2' and 's_3' be abbreviations respectively for the sentences '*S* will be ϕ', 'S is ϕ' and 'S was ϕ'. The following entailment rules bring out what is entailed by a statement endorsing a present use of one of these sentences.

(R_1) That the statement made by now using a sentence synonymous with 's_2' is true entails that the statement made by the use at any time earlier than now of a sentence synonymous with 's_1' was true and that the statement made by the use at any time later than now of a sentence synonymous with 's_3' will be true.

(R_2) That the statement made by now using a sentence synonymous with 's_1' is true entails that the statement made by the use at any time earlier than now of a sentence synonymous with 's_1' was true and that the statement made by the use at any time which is later than now and not later than *S*'s being ϕ of a sentence synonymous with 's_2' will be true and that the statement made by the use at any time which is neither earlier than now nor between now and '*S*'s being ϕ nor simultaneous with *S*'s being ϕ of a sentence synonymous with 's_3' will be true.

(R_3) That the statement made by now using a sentence synonymous with 's_3' is true entails that the statement made by the use at any time later than now of a sentence synonymous with 's_3' will be true and that the statement made by the use at any time which is neither earlier than *S*'s being ϕ nor later than now nor between now and *S*'s being ϕ of a sentence synonymous with 's_2' was true and that the statement made by the use at any time which is neither simultaneous with *S*'s being ϕ nor between now and *S*'s being ϕ nor later than now of a sentence synonymous with 's_1' makes a true statement.

No attempt will be made here to list the entailments of an endorsement of a statement made through a past or future use of an *A*-sentence. These rules, in conjunction with rules (R1)–(R3), could serve as the basis for a genuine tensed or *A*-logic and could easily be supplemented so as to incorporate metrical notions. A genuine tensed logic is one in which entailment relations hold between *A*-statements. The only way in which an *A*-statement can be identified is through the use of an *A*-expression in identifying an occasion on which a certain type of *A*-sentence is used; this is made quite apparent in the way in which the *A*-statements in the antecedents and consequents of the above entailments are identified.[1] One cannot identify an *A*-statement by merely locating through the use of a date an occasion on which an *A*-sentence is used: the reason why the endorsing statement, 'The statement made by the use at t_7 of a sentence synonymous with "It is now raining" *is* true', does not entail any specific *A*-statement is because it fails to identify an *A*-statement. The statement, 'The statement made by now using a sentence synonymous with "It is now raining" is true', does identify an *A*-statement and as a result entails a specific *A*-statement—that it is now raining.

In addition to bringing out the timelessness of truth, rules (*R*1)–(*R*3) also can be used to show the inadequacy of any analysis of 'true' in terms of epistemic terms, as for example the claim that that it is true that *s* entails that relative to evidence now available it is warrantedly assertible that *s*. Rules (*R*1) can be reformulated in terms of 'that'-clauses instead of sentences within inverted commas as follows: 'That is is now true that *S* is ϕ entails that it was true

[1] Thus in the order of identification there is a priority of speech acts over statements, but whether this reflects a priority in the order of Being of speech acts over statements is a mooted point.

that *S* would be ϕ and that it will be true that *S* was ϕ.' That relative to evidence now available it is warrantedly assertible that *S* is ϕ does not entail either that it was warrantedly assertible relative to evidence then available that *S* would be ϕ or that it will be warrantedly assertible relative to evidence which will then be available that *S* was ϕ. It is certainly logically possible that the evidence which is now available indicates that *S* is ϕ but that the evidence which was (will be) available in the past (future) does not indicate that *S* would be (was) ϕ. This shows that an endorsing statement does not entail that there is evidence for the statement being endorsed.

(*b*) While an endorsing statement does not entail that there is evidence for the statement being endorsed, it will be contended that the making of this statement pragmatically or contextually implies that there is such evidence. And it will also be shown that the tensed use of 'true' in metametalinguistic statements depends on this pragmatic implication of the use of endorsing sentences. The first task is to say what will be meant by 'pragmatic implication'.

The queer thing about pragmatic implication is that it is not what someone says—the statement he makes—that pragmatically implies some other statement but his saying it—his using a sentence in a certain context. More precisely:

> (*P*) The use of sentence *s* in context *c* pragmatically implies that *q* if, and only if, the use of *s* in *c* enables the hearer to infer that it is most likely the case that *q* and *s* is usually used in *c* when it is the case that *q* and there is a convention that a speaker should not use *s* in *c* unless it is the case that *q* and if this convention were to be publicly disavowed by the speaker it would frustrate the achievement of the perlocutionary acts for which *s* is usually used in *c*.[1]

This definition can best be clarified by reference to a specific case of pragmatic implication. The use of any declarative sentence *s* in a fact-stating context *c* pragmatically implies that the person who uses *s* believes that *s*. A 'fact-stating' context involves features such as: the speaker is not an actor in a play, is not winking or talking in an ironic tone of voice, etc. The speaker's use of *s* in *c*

[1] For an account of perlocutionary acts see John Austin, *How To Do Things With Words* (Oxford, 1962).

does enable the hearer to infer that it is most likely the case that the speaker believes that *s*, and it is usually the case that a speaker believes that *s* when he uses *s* in *c*. Also there is a convention that a person should not use *s* in *c* when it is not the case that he believes that *s*. Violation of this convention can bring the serious reprimand of being devious and insincere. Furthermore, if a speaker were to publicly disavow this convention, as for example by saying '*s* but I do not believe that *s*', he would frustrate the achievement of the perlocutionary acts for which *s* is usually used in *c*, which include informing, warning, etc.[1] If the hearer knows that the speaker says things that he does not believe, the hearer would discount the things said by the speaker; and as a result the speaker's use of *s* in *c* would not succeed in accomplishing the perlocutionary acts of informing and warning the hearer that *s*.

My analysis of pragmatic implication is very rough, especially in regard to the nature of the convention which underlies such an implication. Rather than trying to further refine my analysis, it will now be applied to the use of endorsing sentences. For any sentence *s* that is used to make a statement for which evidence can be given, a person's use of the sentence 'It is true that *s*' in a fact-stating context pragmatically implies that this person has what is reasonable to take as evidence that that *s* is warrantedly assertible. The reason for restricting this pragmatic implication to the endorsing of statements for which evidence can be given is that not every statement that is endorsed can be supported by evidence, examples of which are 'God exists' (asserted by certain Protestants) and 'Two and two *is* four'. What is *reasonable* to take as evidence that a statement is true or warrantedly assertible is relative to the common-sense and scientific beliefs of a particular age. The evidence possessed by an endorser of a statement need be nothing more than that some authority, such as a newspaper or a scientist, made this statement; but the endorser must be able to give some reasons why this 'authority' is an authority.

It seems to me that it is usually the case that people endorse a

[1] 'Inform' and 'warn' are ambiguous, admitting of an illocutionary and perlocutionary meaning. I can perform the illocutionary act of warning someone by saying, 'I hereby warn you that *s*', but not succeed in performing the perlocutionary act of warning him—getting him to be on his guard—since he might not take my warning seriously.

statement (for which evidence can be given) when they have sufficient evidence to justify their endorsement. In fact, people usually do not use a declarative sentence in a fact-stating context when they do not possess any evidence for what they say, assuming that evidence could be given. This pragmatic implication of the use of a declarative sentence is cancelled by placing an 'I believe' or 'I think' before the declarative sentence. The difference between the use of the declarative sentence '*s*' and the use of 'It is true that *s*' is that the former pragmatically implies only that the speaker has some evidence for that *s* while the latter pragmatically implies that the speaker has sufficient evidence to make that *s* warrantedly assertible. 'I know that *s*' implies, and not just pragmatically implies, that I have sufficient evidence that *s*. Because of this statistical regularity between the use of endorsing sentences and the possession of sufficient evidence by the endorser, a person's use of 'It is true that *s*' in a fact-stating context enables the hearer to infer that it is most likely the case that the speaker has sufficient evidence that *s*. Furthermore, there is a convention that one should use an endorsing sentence in a fact-stating context only when he has sufficient evidence to justify his endorsement: violation of this convention leaves one open to the charge of being reckless and misleading. This convention is of the sort that cannot be openly disavowed without frustrating the achievement of the perlocutionary acts for which endorsing sentences are generally used, these being the same as the perlocutionary acts for which declarative sentences are generally used—to inform, warn, caution, etc. If the hearer knows that a speaker endorses statements whether or not he has sufficient evidence to warrant his endorsement, he would not be informed, warned, etc., by the speaker's endorsement.

It is because the making of an endorsing statement has the above pragmatic implication that there is a use for tensed predications of 'true' in making a metametalinguistic statement. This metametalinguistic use of 'true' and 'not yet true' indicates whether or not sufficient evidence is available to justify the use of the metalinguistic endorsing sentence. This use of 'true' can best be explained by reference to a paradigm case of a future contingent—a possibility for action. The Admiral of the Fleet is deliberating whether to issue an order to his fleet to engage the enemy or to issue an order to the fleet to withdraw. Let us say that if he issues the former order he does *O* and if he issues the latter order he does not-*O*. Let us assume that conditions are such that if he does *O* there will be a sea battle tomorrow and if he does not-*O* there will not be a sea battle tomorrow. The sea battle tomorrow is a future contingent since whether or not it happens depends on how an agent—the Admiral—will choose or decide. Imagine that the Admiral has not yet made up his mind and is being interviewed by reporters. He is asked whether it is true that there will be a sea battle tomorrow. His reply very likely would be, 'It is not yet true that there will be a sea battle tomorrow' or 'It is not yet a fact that there will be a sea battle tomorrow'.

The Admiral's statement can be analysed as follows: There is not now sufficient evidence to justify the use of the endorsing sentence 'It is true that there will be a sea battle tomorrow'. In other words, to use this endorsing sentence now would violate the convention, which underlies the pragmatic implication carried by the use of this endorsing sentence, that one should endorse a statement only when he has sufficient evidence. The reason why the Admiral at this time would consider the evidence to be insufficient is that he is then still deliberating about whether to do *O*. As seen in the preceding chapter, it is logically impossible for an agent to deliberate about doing *X* and at the same time know that he is going to do *X* (not-*X*) or know of the occurrence of an event of which *X* is a necessary cause. The Admiral's doing *O* is a necessary cause of the sea battle tomorrow. Therefore it is logically impossible for him to know of the occurrence of this sea battle while he is still deliberating about whether to do *O*. While he is deliberating he is not concerned with evidence for the occurrence of his doing *O* or the sea battle (although he might be concerned with evidence that he can do *O* or that *O* can bring about the sea battle), but with reasons for doing *O* so as to bring about this battle. Therefore, while he is deliberating he would not have what he would consider to be sufficient evidence to warrant the use of the endorsing sentence, 'It is true that there will be a sea battle tomorrow.'

Someone for whom tomorrow's sea battle is not a possibility for action, such as a common seaman, might have sufficient evidence, based on the Admiral's character, past performance and the present situation, to make it warrantedly assertible that there will be a sea battle tomorrow. The seaman is not an agent in respect to this future event since it is not within his power to perform an intentional action so as to bring about or prevent this battle. He can deliberate about whether to go A.W.O.L., since this is within his power to do, but he cannot deliberate about whether or not there should be a sea battle tomorrow. Since he cannot engage in the latter sort of deliberation there are no limitations placed upon his knowledge of the occurrence or non-occurrence of the sea battle.

While there is no incompatibility between the Admiral's deliberating about whether to bring about or prevent a sea battle tomorrow and some other person knowing what the Admiral will do and as a result whether or not there will be a sea battle, it seems to me that this must necessarily be the exception. Our activities as agents presuppose not only that there are certain aspects of the future which cannot be known by us, but also that these unknowable aspects of the future—the ones over which we are deliberating—are not known as a rule by some person for whom these aspects of the future are not possibilities for action. If another person could usually predict on the basis of general laws what intentional actions I shall perform I would begin to think of myself in a very different way. My deliberations would now seem to be somehow beside the point, and my intentional actions would appear to be more like those of a conscious puppet. It may be that in the future there will be a scientific theory that makes it possible to predict every event, under every description, but if this should come about it would cause a basic shift in our way of conceiving of ourselves and our actions.

The Admiral's deliberation is terminated by his making up his mind—forming an intention to do *O* or not-*O*. If he decides to do *O*, and also knows that he can carry out this intention and that *O* is sufficient in the conditions that then prevail to bring about the sea battle tomorrow, he would have intentional knowledge (as defined on page 125) of the sea battle. He would now have sufficient evidence to warrant endorsing the prediction that there will be a sea battle tomorrow. If he had a mind to, he might summon the reporters and tell them that it is now true (is now a fact) that there will be a sea battle tomorrow.[1] In saying this he is assert-

[1] Even though the Admiral has made up his mind to do *O*, he might hesitate to say out loud that it is true that he will do *O*. For by endorsing this prediction he is putting himself on record or committing himself to doing *O*, such that if he does not do *O* he

ing that whereas before now it was not justifiable, due to a lack of evidence, to endorse the prediction that there will be a sea battle tomorrow, it is now justifiable to do so. Other expressions that serve the same metametalinguistic use as 'It is now true that *s*' are: 'It is already true that *s*', 'It is already decided (determined, settled, set, booked, fixed, rigged) that *s*'. All of these expressions make reference to what some agent has intentionally done. A horse trainer tips off his friend Smith by saying, in a whisper of course, 'The fourth race is fixed: Pretty Girl has been given a "goof ball".' Smith might then tip off his friend Jones (good thing that Jones saved Smith's life by retro-warning him) by saying, 'The fourth is a "boat race". It is already true that Pretty Girl will win.' We might call the metametalinguistic use of 'already true (fixed, decided, etc.)' the 'insider's use' or the 'Broadway use'.

The above metametalinguistic uses of tensed predications of 'true' can help to make sense of the claim of the Temporalistic Theory of Truth that a statement about a future contingent is now neither true nor false but will become one or the other with the passage of time.[1] When the Admiral had not yet made up his mind he answered the reporter's question whether it was true that there would be a sea battle the following day by saying, 'It is not now (yet) true that there will be a sea battle tomorrow'; however, had he instead been asked whether it was true that there would not be a sea battle the following day he would have replied by saying, 'It is not now (yet) true that there will not be a sea battle tomorrow.' Since the Admiral was still deliberating about whether to do *O* or not-*O* he could not have had sufficient evidence to justify endorsing either the prediction that there would be a sea battle the following day or its contradictory. At that time, as far as the Admiral was concerned, it was not yet true that there would be a sea battle the following day and it was not yet true that there would not be a sea battle the following day: in other words,

[1] The question of whether *A*-statements can be rendered by *B*-statements without loss of meaning is totally irrelevant to the sea battle tomorrow problem. The Temporalistic Theory would hold that if the statement made by now using 'There will be a sea battle tomorrow' is neither true nor false now then so too is the statement made by now using the *B*-sentence 'A sea battle *takes place* at t_7', assuming that tomorrow's date is t_7.

could be called to account for it by others. By not declaring out loud his present intention to do *O*, he is reserving the right to change his mind up until the time he actually gives the order to attack or retreat.

it was then neither true nor false that there would be a sea battle the following day. We usually do not say things like 'It is now neither true nor false that *s*' or 'It is now neither true that *s* nor true that not-*s*'. The reason is that we can accomplish the same thing more economically by saying simply 'It is not yet true that *s* (not-*s*)'. This means that relative to available evidence we are not now warranted in either endorsing that *s* or endorsing that not-*s*.

It follows that the use of 'It is now neither true nor false that *s*', like the more common 'It is not yet true that *s* (not-*s*)', makes a metametalinguistic statement. To say that a statement is now neither true nor false is not to say that it has no truth-value or that it has the queer truth-value of 'neuter', 'indeterminate' or '$\frac{1}{2}$' of certain three-valued logics. These logics confound the metametalinguistic use of 'true' with its metalinguistic use, and thereby are led to think that a statement can have a truth-value other than 'true' or 'false'. A more promising way of interpreting the Temporalistic Theory of Truth is as not denying that every statement is either true or false but as denying that every statement is either determinately (necessarily) true or determinately (necessarily) false. The modal notions of determinately or necessarily true make sense when unpacked epistemically in terms of the 'Broadway use' of 'true'. A prediction that *s* is determinately or necessarily true means that it is already true, i.e. it is already decided, settled, fixed, etc., that *s*.

Sense can also be made of the Temporalistic Theory's claim that statements 'become true'. Corresponding to the metalinguistic and metametalinguistic uses of 'true now' are (*a*) a metalinguistic and (*b*) a metametalinguistic use of 'become true'.

(*a*) When 'become true' is used in a metalinguistic endorsing statement it means that, whereas before now the use of the sentence employed in making the statement being endorsed would not have made a true statement, its use now will make a true statement. When *S* becomes ϕ we say that it becomes true that *S* is ϕ, meaning that to use a sentence synonymous with '*S* is ϕ' now makes a true statement, whereas before it did not. Obviously, this metalinguistic use of 'become true', like the metalinguistic use of 'true now', applies only to cases where an *A*-statement is being endorsed.

(*b*) The metametalinguistic use of 'become true' means that, whereas before now it was unwarranted, due to an insufficiency of evidence, to endorse the statement which is said to become true, such an endorsement is now warranted. A statement becomes true in this sense when there becomes available sufficient evidence to warrant endorsing it. When a prediction is of a future choice of some agent or of an event simultaneous with an agent's choice it does not become true until the predicted event occurs. In such a case the prediction that S will be ϕ becomes true when it becomes true in sense (*a*) of 'become true' that S is ϕ.

F. Waismann and others have found insuperable difficulties with this:

> At what moment does the statement 'it will rain tomorrow' begin to be true? When the first drop falls to the ground? And supposing that it will not rain, when will the statement begin to be false? Just at the end of the day, at 12 p.m. sharp? . . . We wouldn't know how to answer these questions; this is due not to any particular ignorance or stupidity on our part but to the fact that something has gone wrong with the way the word 'true' and 'false' are applied here.[1]

Waismann's objection is based on a confounding of the question of the 'open texture' of meaning[2] (i.e. for any event expression there is a conceivable situation in which we would not be sure whether or not to use it) with the question of what is involved in saying that a prediction becomes true. I shall argue that these two questions are independent of each other, that even though every event expression has an open texture of meaning nevertheless we can point to paradigm cases in which it is said that a prediction about the occurrence of a certain type of event 'becomes true'.

The obvious reply to Waismann's objection is that a prediction becomes true when the predicted event occurs or takes place. But, Waismann would ask, at exactly what instant does an event take place? If we can answer this question then we will *ipso facto* be answering the question concerning at exactly what instant a prediction becomes true, for it becomes true at the same instant at which the predicted event occurs. If Waismann's objection proves anything it proves too much, for it would show not only

[1] F. Waismann, 'How I See Philosophy', in *Contemporary British Philosophy*, series three, edited by H. D. Lewis, pp. 256–257.

[2] See Waismann's 'Verifiability', in *Logic and Language*, series one, edited by A. Flew, p. 124.

that it is meaningless to speak of predictions becoming true but also of events taking place, occurring, happening, becoming present, coming to pass, etc. Since we *do* speak of events taking place, etc., there must be something wrong with Waismann's argument. And if we can meaningfully speak of events becoming present or taking place then by the same token we must also be able to speak meaningfully about statements predicting the occurrence of such events as becoming true.

In order to answer Waismann's question concerning exactly when an event takes place (and as a result a prediction about it becomes true), we shall briefly, and in a most incomplete fashion, discuss some of the different types of event expressions and the sense in which the different types of events designated by these expressions can be said to take place or become present. We might classify event expressions into two main classes: (A) expressions which do not refer to a process, and (B) expressions which do. Each of these classes contains two main subdivisions.

Class (A) is subdivided into event expressions which can be predicated only of *single* instants of time (to be called 'achievement expressions') and those which can be predicated for shorter or longer periods of time (to be called 'state expressions').[1]

Achievement expressions refer to events which can be dated but not clocked, such as reaching the finish line, beginning a race, getting the point of a joke or an argument. A statement predicting the occurrence of an achievement event becomes true at the same instant at which the event occurs. A prediction that a certain horse will win a race becomes true or turns out to be correct or is fulfilled at the instant at which the horse reaches the finish line first. In cases of a protest or a photo finish the horse can be said to be the winner only when the judge's decision is announced; only then would the winning tickets be cashed. As a general rule, a prediction becomes true when the bets made on this prediction are paid off. The fact that there are conceivable cases (such as where 100 horses in a steeplechase race finish in a dead heat) which we would not know whether or not to classify as instances of a certain horse winning a race does not mean that there are not paradigm cases of horses winning races and therefore of predictions about horses winning races becoming true.

[1] I am following Zeno Vendler's classification of different types of tensed verbs in his article, 'Verbs and Time', *Philosophical Review*, LXVI (1957).

Whereas achievements involve a unique and definite time instant, dispositional *states* involve time instants in an indefinite and non-unique sense, examples of which are 'knowing how to', 'knowing that', 'loving'. In regard to such states, two sorts of predictions are commonly made—those about an achievement event initiating or terminating a state and those about how long a state lasts. The generic state of being married is initiated by the achievement event of getting married. A prediction about an achievement which initiates a state becomes true at the instant at which the achievement event occurs. A prediction about how long a state will last, such as 'Doris Duke will be married between t_1 and t_4', means that at any instant between t_1 and t_4 Doris Duke will be married. Such a prediction becomes true at t_4, for only then can we be sure, especially in her case, that the prediction is true.

Class (B) is subdivided into event expressions which refer to processes having a definite goal or termination (to be called 'accomplishment expressions') and those which refer to processes having no such goal (to be called 'activity expressions').

Accomplishments are such goal-directed processes as drawing a circle and running a mile. They imply the notion of a unique and definite period of time. A statement predicting the occurrence of an accomplishment process becomes true only at the instant at which the goal or climax is reached. A man cannot be said to draw a circle until it is completed; he might stop in the middle and have drawn only an arc. The climax of an accomplishment process 'casts its shadow backward, giving a new color to all that went before'.[1] The beginning of a process can be said to be the beginning of an accomplishment process only from the vantage point of a later moment of time when the process has reached its goal. Just as being true, to use Ryle's phrase,[2] is a valedictory epithet given to a prediction from a later moment of time when the predicted event occurs, so too being the beginning of a certain type of accomplishment process is a valedictory epithet given to the beginning of a process from a later moment of time when the process reaches its goal, thereby proving itself to be an instance of the type of accomplishment in question. Only at that instant would

[1] *Philosophical Review*, LXVI (1957), p. 146.

[2] Gilbert Ryle, *Dilemmas* (Cambridge, 1954), p. 20.

a statement predicting the beginning (or end) of such an accomplishment process become true. Once again, that any accomplishment expression has an open texture of meaning (e.g. when is a drawn figure an oval rather than a circle?) does not mean that there are not paradigm cases of the use of such expressions.

Activities do not take any definite period of time; they go on in time in a homogeneous way, not being directed to any goal. Activities such as running, raining, walking, a sea battle, call for a period of time that is not unique or definite. Because an activity is not an instantaneous occurrence, like an achievement, we cannot be sure at the instant at which a process begins that it is an instance of a certain type of activity; and since an activity is not directed to a final goal, as is an accomplishment process, we do not know at exactly what instant we are justified in calling this process a bona fide instance of a certain type of activity. Herein we are confronted at the outset with the problem of the open texture of meaning since ordinarily we do not have explicit rules determining our use of activity expressions. Reference to the context will eliminate some, but only some, of the vagueness surrounding an expression's use. For example, if it is predicted that a certain one-year-old will walk tomorrow (for the first time) the parents would probably demand only three steps before they would say that the prediction becomes true, though the next-door neighbours, whose one-year-old has not walked yet, might demand at least five steps. On the other hand, if it is predicted that Dr. Barbara Moore will walk tomorrow we might demand that the old girl go at least fifteen miles before we say that the prediction becomes true, though one of her vegetarian cohorts might be satisfied with only ten miles.

Generally we would say that a prediction about an activity event becomes true at the time at which we are *reasonably* sure that a given process is an activity of the sort predicted. Suppose all prior conditions point to rain; suddenly it starts pouring. At that instant I would be willing to say that a prediction of rain becomes true, though, if it should suddenly stop precipitating a few seconds later, I might retract this by saying that it started as if it was going to rain but stopped. This shows that ordinarily we do not have definite rules stating how long it must precipitate before it can be called rain. There is no reason why the weather department cannot make the rules more precise, in which case the activity expression 'rain' would function like an accomplishment expression, since we could now say at exactly what instant a given process can be called an instance of rain. In conclusion, it is significant that Waismann, in order to prove that a prediction cannot meaningfully be said to become true, used as an example 'rain'—an activity expression with a large open texture of meaning—and neglected all other types of event expressions. Even if his argument directed against activity expressions were sound, it would not show that it is meaningless to speak of predictions about other types of events becoming true.

Since the metametalinguistic use of 'become true' and 'not yet true' makes reference to the state of available evidence it might be thought that a retrodiction (a past tense statement) could also be *not yet true* when there is insufficient evidence for the past event it reports and could *become true* when sufficient evidence is unearthed. It is significant that we do not speak of a retrodiction as 'becoming true' but only as 'turning out to be true', in the sense of being discovered to be true. I suspect that this is due to the fact that a prediction turns out to be true in a more dramatic and definite way than a retrodiction can: a prediction is verified by waiting to directly verify the occurrence or non-occurrence of the predicted event, but a retrodiction can be verified only indirectly by gathering evidence, such as traces, of the retrodicted event. Closely connected with this is the fact that we do not speak of a retrodiction as 'not yet true', even when there is insufficient evidence for it. This is probably due to the fact that there are no past possibilities for action, and as a result our lack of sufficient evidence for a retrodiction can never be due to the fact that we are deliberating about causing or preventing the occurrence of the retrodicted event. In regard to a retrodiction about a sea battle yesterday, what is sufficient evidence to justify a common seaman's endorsing the statement is also sufficient for the Admiral's doing so, but this is not so in regard to a prediction about a sea battle tomorrow.

Another way of formulating the logical asymmetry between the past and future in regard to possibilities for action is as follows:

(*T*) We can now act in such a way so as to falsify or make true a prediction but not a retrodiction.

Making true (Falsifying) a statement does not refer to the act of discovering that this statement is true (false) but to the performance of some intentional action which either is (is incompatible with) the event reported by this statement or brings about (prevents) this reported event. Logical asymmetry (*T*) needs further qualifications so as to meet certain counter-examples, which also are counter-examples to the claim that there are only future possibilities for action.

First we ought to dispose of some alleged counter-examples to (*T*) which are not really counter-examples. One class of such bogus counter-examples is where a retrodictive sentence is used to make a statement about the past and subsequently the meaning of some word(s) in this sentence is altered so that the statement that is now made through the use of this sentence is no longer true. By changing the meaning of some word we are not falsifying the original retrodictive *statement*—what was said through the use of the retrodictive sentence before its meaning was changed. A judge has the power to revise a precedent and can thereby change the criteria for the use of some term. Suppose that it is now stated that Jones was duly tried and convicted of arson. His case is subsequently reviewed and the judge rules that certain evidence introduced at his trial is inadmissible, and, in ruling this type of evidence as inadmissible, is changing a precedent for admissibility of evidence. By doing this the judge is changing the criteria for the use of 'duly tried and convicted', and therefore is not falsifying the original retrodictive statement. Consider a more obvious case of 'falsifying' a retrodictive statement. It is stated that Lepke was electrocuted and we then change the meaning of 'electrocuted' so that it now means 'gassed to death'. Certainly we do not falsify the original retrodictive statement nor do we falsify or make true the new retrodictive statement made through the use of the sentence containing 'electrocuted' with its new meaning.

Another class of bogus counter-examples to (*T*) consists of cases in which a person utters or inscribes a retrodictive sentence token but does not then make a statement but rather makes-a-statement-to-be, i.e. tokens a sentence for the purpose of making a statement when it is seen or heard at some future time. This often happens in broadcasting where shows are taped ahead of time. For example, a radio announcer is making a tape-recording that is to be played at the stroke of midnight on New Year's Eve.

He utters this sentence token, 'Guy Lombardo has just finished playing *Auld Lang Syne*'. (Guy Lombardo has assured the network that he will play this song just before midnight.) Guy Lombardo learns of this tape-recording and is slighted because the announcer is not going to cover his broadcast live. He says to himself: 'I'll fix that s.o.b., I won't play *Auld Lang Syne*.' By desisting from playing *Auld Lang Syne* is Guy Lombardo falsifying the announcer's retrodictive statement? No, because at the time he desists from playing this song the announcer has not yet made his retrodictive statement: that is made at a later time—midnight. Prior to midnight the announcer has only made-a-statement-to-be through his utterance of a retrodictive sentence token: his utterance-token makes a statement only when it is heard at midnight on New Year's Eve. But isn't Guy Lombardo then falsifying the future retrodictive statement made by the announcer? Yes, but this is not a counter-example to (*T*) since it is concerned only with retrodictive and predictive statements made NOW.[1]

Now for some genuine counter-examples to (*T*). One class of such counter-examples involves a retrodiction that is metalinguistic because it contains as one of its parts either a sentence within quotation marks or a noun 'that'-clause. In January, Tom predicts that Harry will vacation in Maine during the month of June. In May, Dick informs Harry of Tom's prediction and adds 'Tom was right' or 'Tom's prediction was true' or 'Tom knew then that you would vacation in Maine during the month of June'. Harry dislikes Dick so much that he purposely does not vacation in Maine during June so as to falsify Dick's retrodiction, 'Tom was right', etc. This class of counter-examples can be met by restricting (*T*) so that it applies only to retrodictions and predictions that are not metalinguistic. This is not a completely *ad hoc* restriction, for the action which falsifies the metalinguistic retrodiction does so only in virtue of its falsifying the predictive statement, 'Harry will vacation in Maine during the month of June', which is referred to by the metalinguistic retrodiction.

Another class of counter-examples contains retrodictions which contain what I shall call a 'temporally impure description',

[1] For a more elaborate treatment of this sort of counter-example see my article on 'Falsifying Retrodictions', *Analysis*, December, 1965, along with John Robison's 'A Note on "Falsifying Retrodictions"', in the same issue.

for short an 'impure description'.[1] Examples of such retrodictions are: 'The man who will become (is to become) the 35th President was born in Boston' and 'The patriots did not labor in vain'; for it was possible for Kennedy to falsify the former and it is possible for some very high government official to act in such a stupid way so as to falsify the latter.[2] This latter case is a counter-example not only to (*T*) but also to the thesis of the preceding chapter that it is impossible for a present event to have a past effect, for by acting as he does now this official causes the patriots to have laboured in vain. What is required is that a precise criterion be developed for pure and impure descriptions. Once this has been accomplished (*T*) can be further restricted so as to exclude retrodictions and predictions containing an impure description.[3]

In developing this criterion we shall have to make use of one unexceptionable assumption and one key terminological distinction. The assumption which we shall make is that if there is a time or period of time there must be something that occurs or obtains at this time or during this period of time. This assumption is a denial of the absolute theory of time, which holds that instants of time are substances, i.e. things which exist independently of other things. Our assumption can be made plausible by reference to the fact that the only way in which we can identify a time or period of time is by reference to some event or state of affairs which happens or endures at this time or period of time.

The terminological distinction we shall make is between 'time-committed-statements' and 'non-time-committed-statements'. A time-committed-statement is one which mentions or refers to some time(s) and which can be true only if the time(s) referred to exist, i.e. there is something which happens or ob-

[1] I am indebted to my colleague, Nicholas Rescher, for this terminology and for many valuable suggestions as to how to develop the distinction between pure and impure descriptions. Edmund Gettier and John Robison also have given me much help in developing this distinction.

[2] 'Labour in vain' can be used to refer to (i) actions which (when successful) do not themselves terminate in the desired result or (ii) actions which (when successful) terminate in the desired result. Our example involves use (i) of this expression.

[3] The distinction between pure and impure descriptions is of vital importance in deontic logic where there are counter-examples employing impure descriptions to many fundamental principles. For instance, one is forbidden to perform an action which entails the occurrence of a forbidden action, and, also, one ought to help someone in distress. But these two principles, each true, seem to clash in the case I describe by saying, 'He is helping the man in distress who robbed the bank.'

tains at the time(s) in question. A non-time-committed-statement is one whose truth does not require that there be some particular time(s). The statement, 'Yesterday someone danced all day', is obviously a time-committed-statement, for it can be true only if there is a yesterday. The denial of this statement, 'There is no one who danced all day yesterday' or 'It is not the case that someone danced all day yesterday', admits of a dual interpretation, one of which is time-committed and the other of which is not. It could mean that there was a yesterday—something took place yesterday—only someone's dancing all day was not among the things which took place then: or it could be interpreted as the disjunctive statement, either there was no yesterday or there was a yesterday but no one danced all day on this day. In the latter interpretation this is a non-time-committed-statement. Usually negative existential statements are time-committed. As has been pointed out numerous times in discussions of negative facts, when it is asserted that something does not exist or happen at some time, it means that something else exists or happens at this time. A hypothetical statement which refers to a time is usually non-time-committed; e.g. 'If someone danced all day yesterday then someone was sweating yesterday', could be true even if there were no yesterday.

In the case of sentences which are used to make time-committed-statements, they can be used to make a statement—say something true or false—only if the time they refer to exists. Thus, if there is no tomorrow, the use of 'There will be a sea battle tomorrow', as well as 'There will not be a sea battle tomorrow', fail to make statements; therefore we are not caught in the embarrassing position of having to say that two contradictory statements are both false. The use of a sentence which makes a non-time-committed-statement does not presuppose the existence of the time referred to by the sentence, if it refers to a time at all.

A. *Impure Descriptive Predicates*. There are two classes of descriptive expressions which must be considered: (A) descriptive predicates, and (B) descriptive referring expressions. We shall begin by trying to formulate a criterion for a pure descriptive predicate, but before doing this it will be necessary to explain what we shall mean by a '*descriptive* predicate', this being a narrower concept than that of a 'predicate'.

A descriptive predicate is, first of all, a predicate: it contains an assertive or propositional word, such as 'is a', and can be joined with a referring expression so as to make a sentence. For example, the predicate 'is ϕ' can be joined with the referring expression 'S' to form the sentence 'S is ϕ'. If the nominalized form of this sentence, which is 'S's being ϕ', can be used in a question of the form, 'When does S's being ϕ occur or obtain?' or 'During what period of times does S's being ϕ obtain?', we shall call the predicate 'is ϕ' a *descriptive* predicate. A predicate of the form 'is earlier than the firing of the canons' is not a descriptive predicate. When this predicate is joined with a referring expression, such as 'Something', it forms the sentence, 'Something is earlier than the firing of the canons', which has as its nominalization, 'Something's being earlier than the firing of the canons'; however, this nominalized expression could not meaningfully be used in a question of the form, 'When does *something's being earlier than the firing of the canons* take place?' A temporal relation between two events is not a further event or state of affairs, and that is why it makes no sense to ask when or during what period of time this relation occurs or obtains. To describe a temporal relation between two events, x and y, is not to describe a third event: to think otherwise is to start on a Bradleian infinite regress. Many predicates containing words for relations will count as descriptive predicates according to our criterion. For instance, 'is taller than John', is a descriptive predicate since when it is joined with a suitable referring expression (e.g. 'Frank'), it makes a sentence about whose nominalization, 'Frank's being taller than John', we can ask when or during what period of time it holds or obtains.

Having clarified what we mean by a 'descriptive predicate', we can now attempt to formulate a criterion for a pure descriptive predicate. The intuitive notion behind an impure descriptive predicate is that it so describes some object or event that an essential reference is made to what happens or does not happen at some time earlier or later than the time at which this object or event has the described property in question. As a first stab at an adequate criterion for a pure descriptive predicate let us consider the following:

(*C*) A pure descriptive predicate is one which when joined with the most general subject term—something or someone—

makes a statement which does not entail any time-committed-statement about a time earlier or later than the reported event.

By the 'reported event' is meant the event referred to by the nominalized version of the statement containing the descriptive predicate which is being tested. The reason why the subject must be the most general subject—something—is that this ensures that the statement formed will not fail to satisfy criterion (*C*) due to the subject expression containing an impure description, as in 'The man who was to become the 35th President was born in Boston'. What causes this statement to fail to satisfy (*C*) is its subject expression, since it can be shown that 'was born in Boston' is a pure descriptive predicate. By requiring that the subject expression be 'something' or 'someone' we ensure that the failure of any statement to satisfy (*C*) will be a result of its predicate expression.

'Is dancing all day today' is a pure descriptive predicate[1] because when it is predicated of 'someone' it makes a statement which satisfies (*C*), that is the statement that someone is dancing all day today does not entail any time-committed-statement about a time earlier or later than the twenty-four hour interval during which the reported event occurs. In other words, this statement is compatible with the statement that there exists no time which is other than today. It is logically possible that the time during which the reported event occurs is the whole of history, there being nothing that occurs or obtains before or after it.[2] In such a case, the Italian folk song, *There's No Tomorrow* (as well as no yesterday), would literally be true. This would mean that there would not be any true time-committed-statement about

[1] We can meaningfully ask, 'When does someone's dancing all day today occur?', to which the answer must be a date or some *B*-statement: it cannot be answered by an *A*-statement or specification of a time-interval, for we already know that the event is present and takes twenty-four hours.

[2] I am assuming that it is not meaningless, as some philosophers have contended, to speak of a 'beginning of time' in the sense of a beginning of the world—a first event. While it is true that in most uses of 'beginning' there is in fact something which happens prior to what begins, this does not show that this is essential to the meaning of 'beginning'. In my opinion, the rules or conventions of ordinary language are not sufficiently precise to legislate either for or against the propriety of speaking of the beginning of time (i.e. the world). Thus, while my use of 'beginning' in speaking of the beginning of time is not sanctioned by ordinary use, neither is it ruled out by ordinary use.

a time earlier or later than today, not even the time-committed-statement that tomorrow it will be the case that someone will have danced all day on the previous day. 'Someone is dancing all day today' does entail the *non*-time-committed-statement, '(t) [if t is a time and is n days earlier (later) than today then it is the case at t that someone is to dance n days later (earlier)],' which statement could be true even if there are no times earlier or later than today.

An impure descriptive predicate is one which, when joined with the most general subject, makes a statement which fails to satisfy (C). The descriptive predicate, 'is dancing the first two days of a four day dance', is impure because when it is predicated of 'someone' makes a statement which violates (C): 'Someone is dancing the first two days of a four day dance' entails a time-committed-statement about what happens at a time later than the time during which the reported event occurs. This statement so describes what takes place during this two-day interval that it entails that something will occur during the following two days—that the person referred to will continue to dance for two more days. If this two-day interval during which the person does the reported dance is the whole of history it cannot be true that this person is doing the first two days of a four-day dance. A predicate containing an intentional term, such as 'is dancing the first two days of an intended four day dance', is not rendered impure on that count, since 'Someone is dancing the first two days of an intended four-day dance' does not entail that this person will continue to dance during the following two days.

Having distinguished between pure and impure descriptive predicates, we are now able to dispose of the counter-example to both the claim that a present event cannot have a past effect and that we cannot now act in such a way so as to falsify or make true a retrodiction, consisting of someone's now acting in such a way so as to cause the patriots to have laboured in vain. 'Labours in vain' is an impure predicate according to (C), since the statement that S labours in vain entails that *subsequent* to S's labouring something will happen or fail to happen which will negate his efforts. In meeting this counter-example it is necessary to restrict these claims respectively to pure descriptions of the cause event and the effect event and to retrodictions containing no impure description.[1]

[1] In designating an event as a cause or effect we are describing it impurely, but what is meant here is that this event must not be further impurely described. A

Unfortunately, there is a fatal weakness in (*C*) consisting in its being too liberal. According to our preanalytic intuitions we would want to classify a descriptive predicate containing a temporal uniqueness condition, such as 'is now dancing the first (last, first and last) polka in history', as impure. The reason is that while this description does not make reference to what happens at times other than the time at which it fits its referent it does make reference to what does *not* happen at these times: someone can now be dancing the first (last) polka in history only if in the past (future) no one dances a polka. But such a descriptive predicate is not impure according to (*C*), since that statement 'Someone is now dancing the first (last) polka in history' does not entail any time-committed-statement about a time earlier or later than the reported event; however, it is incompatible with a statement describing an earlier (later) occurrence of an event fitting exactly the same description, as for example 'Yesterday (Tomorrow) someone danced (will dance) the first (last) polka in history'. This suggests a way in which (*C*) can be revised:

(C_1) A pure descriptive predicate is one which when joined with the most general subject term—something or someone—makes a statement which:

(*a*) does not entail any time-committed-statement about a time earlier or later than the reported event; and

(*b*) is compatible with any statement reporting an earlier or later occurrence of an event fitting exactly the same description.

There are several descriptive predicates which are difficult to classify as pure or impure, since we are not clear as to their correct analysis. An example of such a difficult to classify predicate is one which describes a certain type of speech act, such as promising. If 'He promises' is analysed in such a way that it entails 'There is a previous doubt or problem to which his promise is addressed', then the former statement does entail that something happened prior to his act of promising, and therefore 'promises' is impure. More generally, if one accepts a particular analysis of language

causal statement can be analysed into a statement of the form '*x* causes *y*', and our restriction requires that the event descriptions which are substituted for '*x*' and '*y*' be pure.

such that any use of a linguistic expression logically presupposes previous uses of this expression then any predicate describing a speech act will be impure. It is not our aim to recommend this analysis of 'promise' and of language in general, but only to point out that the problem is not over the correct criterion for distinguishing between pure and impure descriptive predicates but over the proper analysis of certain predicates.

B. *Impure Descriptive Referring Expressions.* A subject expression, as well as a descriptive predicate, can contain an impure description. Assuming that criterion (C_1) for a pure descriptive predicate is correct, we can formulate the criterion for a pure descriptive referring expression as follows:

(C_2) A pure descriptive referring expression is one which when joined with a pure descriptive predicate makes a statement which:

(*a*) does not entail any time-committed-statement about a time earlier or later than the reported event; and

(*b*) is compatible with any statement reporting an earlier or later occurrence of an event fitting exactly the same description.

It is vital that the descriptive referring expression whose purity is being determined be joined with a *pure* descriptive predicate, for if this condition is not fulfilled we could not be sure whether it was the subject or the predicate expression which caused the statement formed to fail to satisfy condition (*a*) and/or (*b*). The same device was used in formulating the criterion for a pure descriptive predicate when it was required that the descriptive predicate being tested be joined with the most general subject so as to ensure that the subject expression is not the source of the impurity, if there is any.

An impure descriptive referring expression is one which when joined with a pure descriptive predicate makes a statement which fails to satisfy condition (*a*) and/or (*b*). 'The man who was to become the 35th President was born in Boston' entails a time-committed-statement about some time later than the reported event—the birth; it violates condition (*a*). Since the predicate, 'was born in Boston', can be shown to be a pure predicate according to (C_1), the source of the impurity must be due to the subject expression. It was shown on page 74 that there often is an ambiguity in the use of a descriptive referring expression, concerning whether or not the referent of this expression is meant to have the described property at the time of the event reported by the statement containing this referring expression. Since the referent of 'The 35th President' is obviously not meant to have this property at the time of his birth 'The 35th President was born in Boston' is elliptical for 'The man who was to become the 35th President was born in Boston'. This does not show that the concept of an impure descriptive referring expression is relative to a sentential context but only that in some of its uses a pure descriptive referring expression, such as 'a President', is elliptical for an impure description—'A person who was to become the President.'

There are many referring expressions about whose analysis we are in doubt. Proper names are particularly troublesome, especially since it is not at all clear that they are even *descriptive* referring expressions. Once again, it is important to stress that the problem is over the correct analysis of these referring expressions and not over the criterion for a pure descriptive referring expression.[1] Assuming that our criteria for pure descriptive predicates and referring expressions are correct, we can now restate thesis (T) so as to meet the two different classes of counter-examples to it:

($T2$) We can now act in such a way so as to falsify or make true a non-metalinguistic prediction which contains no impure

[1] The Humean doctrine of logical atomism cannot be protected against the numerous counter-examples to it by being restricted to statements containing no impure description. The reason is that if the Humean thesis is interpreted to mean that a time-committed-statement about a certain time does not entail a time-committed-statement about an earlier or later time, then his thesis becomes an empty tautology, reducing to the claim that a statement which does not contain any impure descriptions, i.e. a statement which (a) does not entail any time-committed-statement about a time earlier or later than the reported event and (b) . . ., does not entail any time-committed-statement about a time earlier or later than the reported event. Hume's logical atomism must be supported by other means. It might be contended by a Hume or Russell, although not by the Wittgenstein of the *Tractatus*, that the content of our sense experiences can be completely described without recourse to impure descriptions; and therefore an adequate report of what is perceived at one time will never entail a statement about what is the case at an earlier or later time. I do not wish to defend this epistemological form of logical atomism, but merely point out that logical atomism cannot be defended by restricting it to statements which contain no impure descriptions.

descriptions but not a non-metalinguistic retrodiction which contains no impure descriptions.

Even in this restricted form these (T_1) is of the first importance, bringing out the sense in which the future is open and the past closed.

IX

FUTURE INDIVIDUALS

IN this chapter we shall explore yet another sense in which the future is open and the past closed, this consisting in the possibility of *successfully* identifying past (and present) but not future individuals. By a past individual is meant one which existed but no longer exists and by a future individual one which will exist but does not yet exist. That future individuals cannot be *successfully* identified must not be confused with the more common claim that statements about the future, unlike those about the past and present, must be general in logical form. The main difference between these theses, which will not become evident until our discussion is completed, is that the former is based on extralinguistic considerations of an epistemic kind, while the latter is based on purely linguistic considerations concerning the grammatical structure of a sentence and the kind of referring expressions it contains. Before defending the former, we shall discuss the generality-of-predictions thesis, beginning with an unrestricted form of the thesis to the effect that *all* future tense statements must be general in logical form while past and present tense statements can be singular. To meet certain counter-examples this exciting form of the generality-of-predictions thesis will be restricted so as to apply only to predictions whose subject expression refers to a future individual; but even this more modest version will be found wanting.

I. THE UNRESTRICTED FORM OF THE GENERALITY-OF-PREDICTIONS THESIS

Before considering this thesis it will be necessary to distinguish between singular and general statements. It is not helpful to say that a singular, as opposed to a general statement, is *about* an identifiable individual, for a given statement can be 'about', in the sense of its topic of discourse, several things. The principal topic of discourse will depend on the context. For example, 'Ulanova dances' would be principally about the general activity of dancing if it answered the question, 'Does anyone dance?', but it would be principally about the identifiable individual Ulanova if it answered instead the question 'What does Ulanova do for exercise?' However, regardless of what this statement is about, in the sense of its topic of discourse, its subject expression is used to identify the individual Ulanova, and it would therefore count as singular in all of these contexts.

The concept of a general and a singular statement presupposes that a distinction can be made between the logical subject and predicate of certain statements. On pages 72–73 the *logical* subject of a statement was distinguished from a merely *grammatical* subject by the fact that it is controlled by a Strawsonian rule of referring. We might also employ Strawson's 'category criterion' (a predicate term collects other terms in a different way than a subject term does), or some combination of these two criteria.[1] For the sake of simplicity we shall not consider statements that could be said to have more than one logical subject, such as 'Joe Jones will find a dime'. There are at least three different analyses that could be given: (i) The statement has two subject expressions—'Joe Jones' and 'a dime'—since it contains the two-place predicate '____ will find ____'; (ii) 'Joe Jones' is the subject, since the statement contains the predicate '____ will find a dime'; and (iii) 'a dime' is the subject, since it contains the predicate 'Joe Jones will find ____'.[2] If we adopt analysis (i), the statement is singular in respect to the subject 'Joe Jones' and general in respect to the subject 'a dime'; if (ii), it is singular; and if (iii), it is general. To take care of this complication, the unrestricted form of the generality-of-predic-

[1] Cf. Strawson, *Individuals* (London, 1959), and his 'Singular Terms and Predication', *Journal of Philosophy*, LVIII (1961), especially pp. 398–409. Cf. also P. T. Geach, *Reference and Generality* (Ithaca, 1962), pp. 22–80.

[2] See Geach, *op. cit.*, Chapter Two.

tions thesis would have to be rephrased to read: no prediction can contain *a* subject expression that is an identifying or singular expression.

Assuming that a distinction can be made between the subject and predicate of a statement, a singular statement containing only one subject can be defined as one whose subject expression is a *singular identifying expression*. A general statement is one whose subject expression is not of this kind. This, I believe, is how the expressions 'singular and general statement' have been used by most philosophers. The nub of the problem concerns what a singular identifying expression is and what sort of referring expressions qualify as a singular identifying expression. Certainly a necessary condition for being a singular identifying expression is that an expression can be used in some contexts to answer the question, 'Who (Which one) are you talking about?'; whether this is also a sufficient condition will not be determined now.

With these preliminaries out of the way, we can now consider the formulation of this thesis by various philosophers. C. S. Peirce is usually interpreted as a proponent of the thesis that no prediction can be singular. In a famous passage, Peirce considers statements on the following topics: (i) whether prevailing influences upon the theatre will lead to a preponderance of female stars in the immediate future; (ii) concerning New Yorkers who will end their lives the year after next; (iii) 'that of all possible throws of a pair of dice one thirty-sixth will show sixes'. Peirce claims that, whereas we can name famous actors and actresses of former years, we 'can no longer assign proper names' to stars of the immediate future. Regarding (ii), Peirce notes that few, if any, of the New Yorkers who are going to do away with themselves have any idea of their destiny, or any intention to kill themselves two years from now. As for (iii), 'it is impossible so to designate a single one of these possible throws that have not been thrown that the designation shall be applicable to only one definite throw; . . . in their own nature these throws are not individually distinct'. From such examples, Peirce concludes that 'it is only actuality, the force of existence, which bursts the fluidity of the general and produces a discrete unit'.[1]

[1] Peirce, *Collected Papers*, IV (Cambridge, Mass., 1933), # 172. See also Vol. II (1932), ## 146–148, where Peirce maintains that the psychological state of expectation is general, until it is fulfilled by a particular happening; and also Vol. V (1934), # 447.

It is uncertain whether Peirce, in his summing up, means to contrast what is actual with what is merely possible or whether 'actuality' is opposed to what is future. The latter reading, which is required to enlist Peirce as a sponsor of the doctrine that all predictions are general, leaves him vulnerable to many objections. In particular, none of his examples reveal a difference between predictions and retrodictions. Take (i); what prevents us from listing the actresses, by name, whose rise will be influenced by prevailing conditions? The main obstacle, apparently, is our ignorance of the names of up-and-coming actresses. But we may be equally unable to name the great comedians of Plautus' time. Example (ii) fails in a similar way; we simply do not know who is going to commit suicide. Yet there is no reason why psychologists should not devise tests that register a man's suicidal proclivities; then we could give these tests to every New Yorker. This conjecture does suggest some difference between a report of past suicides and a prediction. First, if we do name the prospective victims, this will tip off welfare workers and permit them to take extra measures against the predicted tragedies; naming past suicides does not enable anyone to dissuade people who have already leapt from skyscrapers or blown their brains out. Secondly, a general prediction, 'Eighty-two New Yorkers will kill themselves', does not guide the authorities to the future victims, whereas a list of names allows them to track down the unfortunates. These points do not show that we *must* stick to general predictions of suicides, but only that, if we do not, we make it easier for listeners to falsify our predictions.

Peirce's final illustration is very ambiguous. The sentence, 'of all possible throws . . . one thirty-sixth part will show sixes', could mean either: (iii*a*) 'of possible throws that will be made—that is, of actual future throws—one thirty-sixth part will be double sixes' or (iii*b*) 'of throws that are possible in the future one thirty-sixth part are double sixes'. The difference is crucial, for only (iii*a*) can be used to predict future throws. Thus Peirce is correct in saying, regarding (iii*b*) alone, 'these throws are not individually distinct'. It is logically impossible to actualize all the throws that *might* be made with a pair of dice, for whenever you throw them you forego an infinite number of other throws that could have been made instead. The class of such throws that are possible in the future is infinitely larger than the class of throws that occur in the future. Now consider a group of tosses that we might predict with sentence (iii*a*). A gambler agrees to throw a given pair of dice upon the carpet every ten seconds for a period of five minutes and then destroy the dice. Assume he will carry out his plan; then we have an obvious case where, *pace* Peirce, the future throws of a pair of dice are individually distinct. Each throw the gambler will make is going to occur at a distinguishable time and place. For these reasons we conclude that Peirce's arguments do not support the thesis that all predictions are general.

Of all those who have defended the unrestricted version of the generality-of-predictions thesis, C. D. Broad has made the strongest claims. Reasoning from the premiss that only the present and past are real, Broad asserted that predictions were only pseudo-statements, since there is nothing in the future to which they could correspond or refer. He wrote:

> Something called *tomorrow* is not a constituent of judgments which are grammatically about 'tomorrow', any more than an individual called *Puck* is a constituent of judgments which profess to be about 'Puck'.[1]

For the same reason Broad claimed that no prediction could be true or false; however, it is important to separate the claim that predictions are general from the claim that they are neither true nor false, for these claims are independent of each other. Being a general statement does not debar a statement from having a truth-value, and being singular does not insure that a statement is either true or false.

Along with Peirce and Broad, Ryle is often registered as a defender of the same unrestricted thesis. But notice how Ryle qualifies his position. Arguing from the maxim that you can issue a singular statement only when you have 'got' the item of which you talk, Ryle declares: 'When I predict the next eclipse of the moon, I have indeed got the moon to make statements about, but I have not got her next eclipse to make statements about.' This ruling appears to make 'The moon will undergo an eclipse tomorrow' singular, for we have 'got' the moon now. And that seems to clash with Ryle's sweeping generalization:

[1] Broad, *Scientific Thought*, *op. cit.*, p. 77.

> Statements in the future tense cannot convey singular, but only general propositions, where statements in the present and past tense can convey both. More strictly, a statement to the effect that something will exist or happen is, *in so far*, a general statement.[1]

According to this ruling 'The moon will undergo an eclipse tomorrow' would have to be classified as general, since it is a future tense statement. Do Ryle's seemingly incompatible rules require him to classify a given statement as both singular and general?

We need not speculate about Ryle's position, because Bernard Mayo's essay on 'The Open Future' contains an unqualified endorsement of the generality doctrine, and a stimulating defence of it. He immediately considers a recalcitrant case: the prediction, 'Professor Carnap will fly to the moon', which sounds very much like a statement referring to somebody we have 'got' now. Mayo's response is: 'Although we have "got" the present Professor Carnap, we have not "got" the future Professor Carnap, which is what the sentence is supposed to be about.'[2] This reply derives whatever plausibility it might possess from the aforementioned ambiguity of the preposition 'about' in that in some contexts the prediction could be said to be principally 'about', in the sense of its topic of discourse, Carnap's future or a future flight to the moon. But in all contexts its logical subject expression[3] is used to refer to the identifiable individual Professor Carnap, whom we have 'got' now.[4] Of course no prophet can point at or get his hands on the future event thus foretold—Carnap's lunar voyage—but no subject expression in this prediction refers to that event.

Because the subject expression of 'Professor Carnap will fly to the moon' is used to identify an individual, it must qualify as a

[1] Ryle, *Dilemmas*, *op. cit.*, p. 27.

[2] Mayo, 'The Open Future', *Mind*, LXXI (1962), reprinted in *The Philosophy of Time*, edited by Gale. See A. N. Prior's complaint at Mayo's claim that Prior holds the same view, *Mind*, LXXIII (1964), pp. 268–269. Prior here anticipates some objections we make against Mayo's doctrine.

[3] Mayo and we are analysing this prediction in such a way that it contains the predicate '____ will fly to the moon'. But, as pointed out above, there are other possible analyses of this sort of statement.

[4] Another way of showing that certain predictions are used to identify present individuals is by considering the attitudes and responses that accompany them. When a doctor says to his patient, 'Jones, you will die within six months', it is the present Jones that he is urging to prepare for death. Moreover, it is the present Jones whom we pity and try to comfort—not the future Jones.

singular statement. Unlike the general prediction 'A (Some) man will fly to the moon' it has forestalled the question 'Who?' or 'Which one are you referring to?' Since the singular prediction is used to identify an individual *at the time it is made*, the question, 'Is this the man you *were* referring to?', can be asked of the prophet when the predicted event occurs; for only if Carnap makes it to the moon is the prediction fulfilled. However, if the general prediction is made, this question is not in order because *any* man who makes it to the moon counts as fulfilling the prediction. There is a helpful analogy to be drawn between singular and general predictions, on the one hand, and definite and indefinite promises, on the other.[1] If I make an indefinite promise to give you *a* watch, when I hand a watch over to you in fulfilment of my promise, you cannot ask me if this is the watch I promised; however, had I instead made a definite promise to give you a particular watch, then this question would be in order. Just as any watch will serve as a fulfilment of my indefinite promise, any man who makes it to the moon will fulfil the general prediction, 'A man will land on the moon'. It is irrelevant that the prophet was *thinking of* Carnap when he issued his general prediction, for *thinking of* someone is not the same as *referring to* someone, the latter being a public activity involving a speaker and an audience.

It might be argued that, contrary to appearances, the prophet is referring to a future individual, for only certain future extensions of the present Carnap will count as being identical with him. For instance, the present Carnap might suddenly turn into a mouse immediately after the prediction is made, and this mouse subsequently be placed in a nose cone of a rocket and land on the moon. Would this event count as a fulfilment of the prediction? Our answer will be a result of what criteria we adopt for personal identity. If the mouse has all of Carnap's memories and can still discourse on fine points in modal logic, we might be inclined to say that he is self-identical with the present Carnap, and therefore, that his landing on the moon makes the prediction true. But notice

[1] This example is patterned upon one of Prior's illustrations in 'Identifiable Individuals', *Review of Metaphysics*, 13 (1960). In his earlier treatment of this subject in *Time and Modality* (Oxford, 1957) Prior defended a very different doctrine: herein he claimed that we can name only things that exist now—a view which obliterates asymmetry between past and future. Prior expressed misgivings about this result, saying, 'we are happier about granting that we cannot properly name, and that there are no facts directly about, objects which *do not yet* exist'.

that this puzzle about the identity of Carnap and, thus, about what future events will serve as fulfilling the prediction can arise only because it is assumed that the prophet refers to the present Carnap when he makes his prediction; for the question takes the form of whether some future individual—the mouse—is identical with the *present* Carnap.

A further difficulty for Mayo's thesis is this: if he is right, in spite of all the counter-arguments presented above, in claiming that 'Professor Carnap will fly to the moon' must refer to a future individual whom we have not 'got' now, then to be consistent he would have to claim that the retrodiction 'Professor Carnap was born in 1891' refers to a past individual whom we also have not 'got' now. But then the asymmetry between predictions and retrodictions, which is required by his thesis, is obliterated.

II. A RESTRICTED FORM OF THE GENERALITY-OF-PREDICTIONS THESIS

So far we have argued against the thesis that *all* predictions are general, but it will be noted that all of our counter-examples involved predictions referring to a *present individual*, i.e. an individual that exists now. To meet these counter-examples we shall restrict the generality-of-predictions thesis so that it now reads: Any prediction referring to a future individual must be general, although some retrodictions referring to past individuals are singular; thereby we retain a limited asymmetry between the past and the future. The reason for this limited asymmetry is that a prediction referring to a future individual cannot have a singular identifying expression for its subject. To justify this claim requires that we take a closer look at our ordinary concept of identification, that is, the concept of identification used by police inspectors, witnesses, gem dealers, appraisers of art works, jewel thieves, prospectors, archaeologists, and similarly down-to-earth folk.

Before we examine the various relevant senses of 'identify' so as to be able to determine which kind of referring expressions are singular identifying expressions, it must be pointed out that any attempt to defend the thesis that future individuals cannot be identified, based on our inability to describe and pinpoint future individuals, is doomed to failure. For not only can we give as detailed definite descriptions of future individuals as we can of present and past individuals, but we can always ensure that no more than one individual will satisfy our definite description by incorporating into it spatial and temporal indexical signs that relate our utterance token to the described individual. For example, just as we can describe a past city as 'the city that occupied this spot 4,000 years earlier than now', so too we can ensure uniqueness of reference of a definite description of a future city by saying 'the city that will occupy this spot 4,000 years later than now'. What this shows is that we can pinpoint future individuals with the same accuracy and definiteness as we can present and past individuals. But does this show that we can *identify* future individuals?

Many philosophers count definite descriptions as singular identifying expressions and therefore would answer in the affirmative.[1] A proponent of the limited form of the generality-of-predictions thesis must show that definite descriptions do not count as singular identifying expressions. Here is how he might try to make out a case for this. There are, for our purposes, two relevant senses in which one can identify an individual: (1) by picking it out; and (2) by naming it.

(1) One picks out an individual as the individual who satisfies some description or name. *Webster's New International Dictionary*, second edition (1938), explains this sense of 'to identify' as 'to prove the same (with something described, claimed, or asserted); as, to *identify* stolen property'. Since pointing, indicating, or singling out is so prominent in such accounts as this, we shall call this the ostensive sense of 'identify'. The identificatory use of pronouns is a species of identification in this sense, since pronouns are indexical signs. A woman is said to be able to identify her assailant in the ostensive sense when she is capable of picking him out of a police lineup. Although we can ostensively identify present individuals, it is logically impossible to ostensively identify past and future individuals, since, by definition, they are not around to be pointed at or grabbed by the lapels. There is, however, a weak asymmetry between past and future individuals, for we can pick out past individuals by pointing to their photographs, but there are no photographs of future individuals due to

[1] As an example see Strawson, 'Proper Names', *Proceedings of the Aristotelian Society*, Supp. Vol. 31 (1957), p. 190.

the fact that we can only have a trace of the past.[1] The woman can identify her assailant by picking him out of a rogues' gallery, even though he may have died in the meantime.

(2) One may reveal the identity—perhaps the proper name, code name, or serial number—of some individual. A woman identifies her assailant by giving his name, instead of pointing to him in a lineup or rogues' gallery. During the pause for station identification, the radio announcer says, 'This is WNBC'. For short, we shall call this the *naming* sense of 'identify'. Later on we shall try to show that in the naming sense of 'identify' there is a more striking asymmetry between the past and the future than in the ostensive sense, in that we can identify past, but not future, individuals by giving their names. To make this plausible we shall have to dispose of a number of seeming counter-examples of proper names for future individuals.

Among other legitimate senses of 'identify' we should mention something very much akin to describing: that is, identifying an individual by 'saying what it is', telling the species or class to which it belongs. It is crucial to note, first of all, that this form of identification presupposes that the individual is identifiable in either sense (1) or (2), or both. We must be able to establish, in case our characterization fits more than one thing, just which individual we are characterizing. Presumably that requires picking out or naming. Another feature of describing individuals deserves notice: in practical contexts we often find a sharp contrast between identifying and describing. The assaulted woman may be able to describe her assailant in some detail, even give a uniquely individuating description of him; but if she fails to pick out the man when he appears before her in a police lineup and she cannot give his name, we say that she is unable to identify him. The question, 'What is the identity of the murderer?', cannot be answered by a definite description. Conversely, a student is incapable of describing his lost typewriter; he has forgotten what brand, size, or model it is. If he recognizes it among typewriters at the local pawnshop or if he recalls the manufacturer's serial number, he identifies the machine. Thus, identifying, in both ostensive and naming senses, is quite different from describing. And no

[1] We might get a rough idea of what some unborn child will look like by looking at photographs of his parents and other children already born to them, but these are hardly photographs of this unborn child.

amount of description, definite or indefinite, guarantee identification, in the ostensive or naming sense, of the individual that is described. Consequently, there is no discord between our account of identification and our previous admission that one can give a uniquely individuating *description* of some future individual.

Perhaps we have overlooked some other types of identification, which may work for future individuals. Might we not be able to identify future individuals by giving their titles? In the first place, it can be argued whether giving someone's title, even when this title implies uniqueness, e.g. 'Heavyweight Boxing Champion', counts as identifying him.[1] We shall waive this difficulty. All we need to show is that *if* a title implying uniqueness can count as an identifying expression then no future individual can now be identified through the use of such a title, for the simple reason that a person cannot possess a title until he comes into being. Herein we find an asymmetry between past and future individuals, because dead people have a grip on their titles that future titleholders cannot have. Even when he dies, the Count of Paris has official status as the late Count, although his title passes on to his male progeny. But notice that future individuals who will eventually receive the title have no official status now as 'Future Count of Paris'. The title of 'Saint' is especially interesting because of the asymmetry it reveals between past and future. A title like 'Saint Anthony of Padua' cannot be given to a present individual, however saintly; *a fortiori* future individuals cannot be canonized.

Although one cannot identify future individuals in the ostensive sense, perhaps one can identify them by giving their names. If so, we must retract our limited version of the generality-of-predictions thesis to the effect that predictions whose subject expressions refer to future individuals are never singular, whereas retrodictions referring to past individuals can be singular. What we must show is that we cannot identify future individuals by naming them because they do not have names; whereas past individuals can be identified by name, since a dead man does not lose his name. As a start, notice that towns, libraries, ships and people of the future are seldom mentioned by name. Road maps do not contain the

[1] If I predict, in a certain context, that the President of the United States will attend the banquet and if in the meantime the President has been impeached and a new man takes office, then if the new President attends the banquet my prediction is fulfilled. There are contexts where this would not hold.

names of towns that will be founded. City directories do not list unbuilt edifices. Maritime registers leave out names of vessels yet to be constructed. *Who's Who?* does not contain names of unborn notables.

These and other peculiarities of names favour our thesis, but we cannot afford to neglect a horde of contrary examples. A. J. Ayer describes a difficult case:

> There are parents who are so anxious that *any* son of theirs should go to an exclusive school that they enter the boy's name before he is born and sometimes, if they are optimistic, before he is conceived. The name under which they enter him may fail in its reference, the person to whom it is intended to refer may never be born, but it may succeed. And if it does succeed, the use of the name before the child exists fulfils exactly the same function as its use after the child has come into existence. The risk of its failing in its reference is greater, but this does not make the usage incorrect.[1]

It is misleading for Ayer to speak of 'the boy's name', as if some kind of ownership were possible here. For names, like titles, are issued in the course of some ceremony that the recipient usually attends, and, if not in attendance, is readily at hand. Most children are not present when their parents fill out their birth certificate in the hospital, but at least they are asleep in the nursery with a name tag on their wrist, so that there can be no doubt as to whom is getting the name. A man may not attend a ceremony when he gets the title of Dean, but at least the postman hands him a registered letter, which is a surrogate for his presence at the titling ceremony. Since future individuals cannot participate in current or bygone christenings, initiations, rites of passage, baptisms, or other naming rituals, or have been readily at hand while these ceremonies went on, these future individuals do not yet own the names they will have. Individuals that no longer exist still possess their names, because they did take part in a naming ceremony. One's name, unlike one's title of nobility or one's estate, is not surrendered to heirs or tax collectors when one dies. So, compared with present and past individuals, a future individual has a very weak hold upon a proper name. Thus, whereas we can identify past individuals by names they possess, we cannot identify future individuals by the names they possess.

[1] Ayer, *The Concept of a Person, and Other Essays* (New York, 1963), pp. 249–250.

Ayer overlooks many crucial differences between the logic of a name that is made up *for* a future individual and that of a name *of* a present or past individual. It is quite true that a couple can make up a name 'Max Jones' *for* a future son and can even have this name put on a list of incoming pupils at Eton for some future year. However, 'Max Jones' is not yet the name *of* a boy, because it has not yet been bestowed upon him in a naming ceremony. Therefore, this name can serve to enroll *any* future son of theirs whom they decide to give the name 'Max Jones' to, even one they adopt. How different things are when 'Max Jones' is the name *of* a particular boy. Imagine that the Jones family has two boys, Max and Sam; Max alone appears to have scholarly promise, so only his name is put down for Eton. Then Sam begins to show marks of genius. Can the Joneses decide to switch the names of their sons, so that the more talented one shall attend Eton? School authorities would point out, if they are pedantic, that the boy who was called 'Max' at the time of the enrolment is still the only Jones child registered for Eton. The name 'Max Jones' was, after all, nothing but a device for signing up its bearer. The moral of this fiction should be obvious enough; when you are dealing with present and past individuals, you are not permitted to decide who shall be the bearer of a given name. Only one person will do, just as only one particular watch would serve as a fulfilment of my definite promise. We must conclude that Ayer is mistaken in asserting that 'the use of the name before the child exists fulfils exactly the same function as its use after the child has come into existence'.

But we cannot leave the subject without analysing the dissimilarities between names of present and past individuals and names made up for future individuals. Our very manner of speaking about future individuals and the names we have made up for them suggests a difference. Prospective parents, however sanguine, cannot say that their next son *is* 'Max', as they do when discussing a child who has been born; they say either, 'We are going to call him "Max",' or 'His name will be "Max"'. Another idiomatic difference is that you can say, '"Max" is the name *of* my son' only after your child is born; prior to his birth all you can say is, '"Max" is the name I've chosen *for* my next son'. Furthermore, notice that, if a son should not be born to the Joneses, they could answer the question 'What boy is "Max Jones" a name *for*?', the answer being, 'The son we hoped would be born to us'; but they cannot answer the question 'What boy is "Max Jones" the name *of*?'

In many cases where we make up a name *for* a future individual, such as a building, the grammatical proper name we select functions not only as a name *for* a future building but also as a name *of* a present project, consisting of a set of blueprints, a site, building materials purchased, labour contracts, etc. When the building comes into being a grammatical proper name changes in its logical role from being the name *of a project* and a name *for a building* to being the name *of a building*. Does it sound odd to assert that the same name is at one time the name for a future building, as well as the name of a project, and then becomes the name of a completed building? There are analogies to this change in the nonlinguistic world. A Carpenter sets out to make a door. He begins by carving a handle. Until the rest of the door is completed, this is merely the handle *for* a door. When the handle is attached, it finally becomes the handle *of* a door. This dual usage of names is also found outside discussion of future children and buildings. Names *for* future storms are made up in advance, and become the name *of* a particular storm when the storm occurs, the order of assignment of names being alphabetical. Military and scientific circles use code names to identify projects during the preparatory stages, and after the projects have been carried out the same names serve to identify the completed operations. A combat exercise of May 1964 was called 'Operation Desert Strike'; before that time, this was only the name of a plan. Names like 'Project Vela', 'Project Gnome' and 'Project Shoal' were used first to identify preparations for nuclear tests and afterwards to identify completed experiments.

Once a proper name has been bestowed upon an individual—has become the name *of* an individual—it can be used with the logical force of a proper name in statements referring to any time, even times prior to the time at which the name was bestowed. In this sense alone are proper names tenseless. Peter Geach presented an argument to show that proper names are tenseless in some stronger sense:

> No reference to time is involved in the question whether a proper name in a given use . . . has a bearer. Thus, the proper noun 'Augustus' as used in Roman history books has Octavian for its

> bearer; this is true without temporal qualifications, even though Octavian lived for years before being called by the name; it would be absurd to object to the question 'When was Augustus born?' because the name was not conferred on him then.[1]

In the first place, 'Augustus' is a title that was adopted by all succeeding emperors of Rome. Rather, we shall concentrate on the use of 'Gaius Octavian', which *is* a proper name, and shall try to determine whether its use is subject to temporal distinctions. All Geach's example establishes is that *once* a proper name has been conferred upon an individual it can be used with its normal force as the name *of* an individual in statements referring to any time, even a time prior to the time at which the name was conferred. However, it does nothing to show that a name can be used as a name *of* an individual in statements made prior to the time at which this name is conferred. In this connection it is vital to distinguish between (*a*) the time reference of a statement and (*b*) the time at which the statement is made. Before Octavian came into being and had the name 'Octavian' bestowed upon him, we could not say things like 'Octavian will be conceived by Atia next Thursday evening' or 'If Atia has a miscarriage Octavian will not be born', though we could have said '*A* son will be conceived by Atia next week who will be called "Octavian"'. Of course *after* Octavian has been born and 'Octavian' has become the name *of* Octavian, we can say 'Octavian was conceived at such and such a time'.

Our claim that a name shifts in its logical role from being the name *for* to being the name *of* an individual when that individual comes into being, requires that we have some criterion for saying when an individual comes into being. This problem is exactly parallel to that of determining when a prediction 'becomes true', which requires that we have some criterion for determining when the predicted event occurs. This is hard to formulate. Has an ocean liner come into existence when its keel is laid? Must the hull be finished? What if the engines are not installed yet? Does it need to be furnished before it is a thing of the present, upon which we can bestow a name? With regard to human beings, the questions are both trickier and more significant from a moral, legal and religious standpoint. Is a foetus a person? We can identify it in the demonstrative sense by pointing. For purposes of naming, one

[1] Geach, *op. cit.*, p. 29.

could perform a baptismal ceremony. If water is considered necessary, it is possible to insert a tiny hose into the mother's body and spray the foetus while naming it. By this practice we might be able to persuade strict Catholics to save the mother rather than the baby in difficult childbirths.

Legal thought concerning liability towards prenatal children contains interesting standards of individuality. In the case of *Dietrich v. Northampton*, decided by the Supreme Court of Massachusetts in 1884, the majority ruled that if a pregnant woman is injured through another person's negligence and if the injury causes her child to be born maimed or disabled, the child cannot sue for damages. Speaking through Justice Holmes, the court held that the child is 'part of the mother's body', and therefore not a distinct individual ((1884) 138 Mass. 14). This precedent was reinforced by the decision of the New York Court of Appeals in the case of *Drobner v. Peters*, which rules that Drobner was 'not a person but a part of the body of his mother. . . . As the injury was to his mother he has no cause of action' ((1921) 232 NY 220–221). In 1951, the New York Court of Appeals reversed precedent when it decided for plaintiff in the case of *Woods v. Lancet*. Judge Desmond's opinion for the majority was founded on a quite different standard of individuality, namely that 'an unborn viable child is capable of independent existence and hence should be regarded as a separate entity' ((1951) 303 NY 349).[1]

Fortunately we do not have to defend either of these competing criteria for saying that an individual has come into being. Regardless of what criteria one adopts for saying when an individual comes into existence—conception, seventh month of pregnancy, or birth—one must recognize a shift in the logical force of the name *for* this individual when it comes into being (whenever that might be); for only then does it become the name *of* this individual. Once again, we must not confound the problem of the 'open texture' of meaning with what it means to say that some individual begins to exist, and as a result the name *for* him shifts to being a name *of* him.

The preceding has presented the strongest case that can be made out for the restricted form of the generality-of-predictions thesis; however, I do not believe that it succeeds in showing that a

[1] Adapted from C. Gordon Post, *An Introduction to the Law* (Englewood-Cliffs, 1963), pp. 89–91.

prediction referring to a future individual cannot have a singular identifying expression for its subject. There are at least two serious objections that can be raised against this thesis. First, contrary to what this thesis claims, definite or uniquely individuating descriptions are singular identifying expressions. Two different justifications can be given for classifying them in this manner.

(*a*) Definite descriptions, in some contexts, can serve to answer or forestall a question of the form, 'Who (Which one) are you talking about?' There is no one type of referring expression that alone can successfully answer such a question in all contexts. Consider a context in which nothing is known about the man who lives next door, not even his name:

'Who robbed the store?'
'The man next door.'
'Who is the man next door?'
'Max Jones.'

But in a different context in which the referent of 'Max Jones' is not known but a good deal is known about the man next door the conversation might go differently:

'Who robbed the store?'
'Max Jones.'
'Who is Max Jones?'
'The man next door.'

As is well known, ostensive identification often can fail through ambiguity, e.g. 'Which one of these men are you pointing at?' Proper names, definite descriptions and demonstratives mutually support each other.

That a referring expression can serve to answer or forestall a question of the form 'Who (Which one)?' should be both a necessary and a sufficient condition for being a singular identifying expression. Our ordinary use of 'identify' requires this. It may be that in certain contexts only one type of referring expression will serve to identify; for example, in the police lineup case a demonstrative was required, while only a proper name would serve to answer the question, 'What is the identity of the killer?' But this is compatible with definite descriptions having an identificatory function in certain contexts. Since definite descriptions are singular identifying expressions, the prediction, 'The first son who will be born to the Joneses will go to Eton' is singular and thereby constitutes a counter-example to the thesis that a prediction referring to a future individual cannot have a singular identifying expression as its subject.

(*b*) A proper name requires a *backing* in definite descriptions. This is not to say that a proper name has a connotation; for in regard to a proper name there is no one definite description which is necessary nor set of definite descriptions which together is sufficient for its correct application. But what is required for a speaker to succeed in identifying some individual through the use of a proper name is that the majority of definite descriptions which he believes to characterize the referent of the name coincide with the majority of definite descriptions that the hearer believes to fit its referent. Proper names have a practical utility: it would be cumbersome to cash them in on every occasion in favour of a set of definite descriptions by which the speaker and hearer back up the use of the name.[1] Since proper names require a backing—must be cashable—in terms of definite descriptions, and since proper names serve an identificatory function, definite descriptions should also serve to identify.

A second objection, also directed towards showing that a prediction referring to a future individual can have a singular identifying expression for its subject, is based on the possibility of now bestowing a proper name on a future individual. The laws and conventions of our society do not permit us to confer or bestow a proper name on a future individual, but they could easily be altered so as to permit this. The Joneses then would be allowed to fill out their future son's birth certificate before he is conceived and could even baptize him *in absentia* by sprinkling the crib that he will eventually sleep in with water. If we can give a proper name *to* a future individual, we can have a proper name *of*, and not just *for*, a future individual. Assuming that 'Max Jones' is the name the Joneses have given to their future son, the prediction 'Max Jones will go to Eton' is singular.

III. THE IMPOSSIBILITY OF *SUCCESSFULLY* IDENTIFYING FUTURE INDIVIDUALS

I believe that the restricted form of the generality-of-predictions thesis can be revised so as to work. The restricted, as well as the

[1] This view of proper names is espoused by John Searle in 'Proper Names', *Mind*, LXVII (1958).

unrestricted, thesis is based on purely linguistic considerations concerning the kind of referring expression that can occur as the subject of a prediction referring to a future individual. Let us grant that the above two objections succeed in showing, contrary to this thesis, that such a prediction can have a singular identifying expression for its subject. This does not establish that such a prediction can *succeed* in identifying a future individual. It is one thing for an expression to be a singular identifying expression and another for its use on some occasion to succeed in identifying a particular individual. In saying that an expression is a singular identifying expression we are referring to its intended function and not thereby saying that it is successful in achieving that function, just as to say of a play that it is a comedy is not thereby to say that it is comical: there is no more a contradiction in saying that a specific use of an identifying expression fails to identify than in saying that a certain comedy is not funny.

The thesis I shall defend is that we cannot *successfully* use a singular identifying expression—a proper name, definite description or demonstrative—to identify a future individual. In defence of this it will be shown, first, that the criterion for a successful use of an identifying expression is based on extra-linguistic considerations of an epistemic sort; and, second, that, due to the logical asymmetries of the preceding two chapters, this epistemic criterion for successful identification is not satisfied by our use of singular identifying expressions for future individuals.

The following remarks about successful identification will apply only to spatio-temporal individuals. The criterion for the successful use of a singular identifying expression is:

(*S*) A singular identifying expression is successfully used to identify a spatio-temporal individual if, and only if, there exists in the spatio-temporal order a referent for the use of this expression, and both the speaker and hearer know what or who this referent is.

It is obvious that a proper name, definite description or demonstrative can be successfully used to identify a spatio-temporal individual only if it succeeds in referring to a spatio-temporal particular, and moreover one which is known to both the speaker and hearer. Whether the use of a singular identifying expression satisfies (*S*) cannot be determined by linguistic considerations; for no amount of knowledge of vocabulary or syntax can determine whether it has a referent, and moreover one that is known by the speaker and hearer.

There can be no doubt that we often succeed in identifying past and present individuals through the use of singular identifying expressions. For instance, the use of the definite description 'the man who wrote *Principia Ethica*' or the proper name 'G. E. Moore' does succeed, according to (*S*), in identifying a past individual: for it is the case that it has a referent, and moreover one whom we know in the sense that we know many facts about him other than the fact that he wrote *Principia Ethica* or that he was called 'G. E. Moore'. Can we ever know that a specific use of a singular identifying expression for a future individual satisfies (*S*)? I think not. In the first place, there is always a genuine doubt whether there will exist a referent for the use of a singular identifying expression for a future individual. This is due to the fact that we cannot have a trace of the future, only the past, and that many aspects of the future are unknowable to us because we are agents who deliberate and perform intentional actions as a result of these deliberations. There is no doubt that 'The Metropolitan Opera House' refers to a past building, for we have traces, in the form of photographs and memories, of it; but in respect to a name for a future opera house there is a genuine doubt whether it will have a referent. Not only can we have no non-inferential knowledge of its existence, but also we may *decide* to junk the project because it is too costly. 'Max Jones' may fail to refer to a future son of the Joneses not only because of many medical contingencies, such as miscarriages and still-births, but also because the Joneses may *decide* not to conceive or, if too late for that, to have an abortion, etc.

Not only is there a genuine doubt whether the use of a singular identifying expression for a future individual will have a referent, but also this referent is not known to the speaker and hearer, in that they know no categorical facts about this referent. Even if we somehow knew that the name 'Max Jones' will have a future individual for its referent, we would not know any facts about this individual other than that his name is 'Max Jones'. Of course the Joneses could tell us that Max Jones will go to Eton, will play on the soccer team, and the like, but all of these statements are doubtful since he may be a moron, poorly co-ordinated, and the like. Their imaginative flights of fancy do not answer or forestall the question 'Who (Which one) are you talking about?' In an epistemic sense, a future individual is not yet a part of our world. There can be no records of him, and therefore his history cannot be written yet. We can only make up stories about future individuals. The attempt of the Joneses to put some flesh on to the referent of 'Max Jones' resembles a fictional narration.

There are good epistemic reasons why we do not permit a name to be given *to* (or bestowed *on*) a future individual, and as a result permit there to be only names *for*, and not *of*, future individuals. 'To give a name *to*' and 'to be a name *of*' are success expressions in that both that the name '*S*' is given *to* x and that '*S*' is the name *of* x entail that x exists.[1] Because these verbs take an objective accusative it would, for the above reasons, be epistemically too risky to give a name *to* a future individual, and to permit there to be a name *of* a future individual. Notice the ease with which we can give a name *to* a past individual, since at least we now have traces of his existence and labours.[2]

[1] It might be thought that the use of proper names in fiction is a counter-example to this, but the reader is reminded that my discussion applies only to the identification of *spatio-temporal* individuals. Even without this restriction, I doubt that fictional proper names would constitute a counter-example. Fictional proper names are not names *of*, but only names *for* characters of the author's own creation. Such a name is not bestowed *on* or given *to* some person that the author comes upon. Demonstratives are never used in the assignment of a fictional name. This brings out the importance sense in which the language of fiction is severed from the world of experience.

[2] Much of this chapter has been drawn from an article I co-authored with Irving Thalberg—'The Generality of Predictions', *Journal of Philosophy*, LXII (1965). In this article we followed Prior's article on 'Identifiable Individuals', *op. cit.*, in defending the restricted form of the generality-of-predictions thesis. I do not know whether Professor Thalberg would agree with my criticisms of this thesis and the manner in which I revised it.

Part Four

THE OBJECTIVITY OF TEMPORAL BECOMING

X

A PRIORI ARGUMENTS FOR THE SUBJECTIVITY OF BECOMING

IS temporal becoming, for short, becoming, in some sense 'subjective'? Would events have *A*-determinations and change in respect to them even in a world devoid of subjects or persons? None of the tenets argued for in the preceding chapters are relevant to this question. For example, the conclusion of Chapter IV concerning the ineliminability of *A*-determinations does not establish that *A*-determinations are objective features of events; for the fact, if it is a fact, that psychological terms cannot be reduced to physical terms certainly does not establish that psychological events and states could exist or occur in a world devoid of subjects. Nor do any of the logical asymmetries between past and future of Part Three indicate that there is an objective present or now in nature. In particular these asymmetries do not show that there is a unique present: for it is true of *every* event in history that actions performed at this time could not cause an event earlier than this time, that persons existing at this time could deliberate and choose only in respect to events later than that time, and that individuals existing at an earlier, but not a later, time could then be *successfully* identified. In brief, none of these logical asymmetries can serve as a criterion for determining what event(s) or time is the present.

My discussion of the question concerning the subjectivity of becoming will concentrate on the status of the present. Since events are past and future only in relation to the present or now, if there should be something subjective about the determination of an event as present, then the determination of an event as past or future will also be subjective. Also, if the determination of an event as present is subjective then the above logical asymmetries are reduced to the status of criteria for distinguishing between an earlier and later 'direction' of time. Thus everything hinges on the status of the present.

If it can be shown that the present is an objective feature of events it will be shown that the *A*-Series is objective. But does this also show that becoming is objective? Becoming requires that there be a sequence of *A*-Series, i.e. that events change in respect to their *A*-determinations. How can this series of *A*-Series be derived from a single *A*-Series? It can easily be shown that if there is one *A*-Series there must be a series of *A*-Series. Assume that the *A*-Series consists of events *M*, *N* and *O*, which are respectively past, present and future. A past (future) event by definition is one which was (will be) present. Therefore, there was (will be) an *A*-Series in which *M* (*O*) is present. Thus, if there is one *A*-Series there is becoming—a series of *A*-Series: and if the *A*-Series is objective so too is becoming. Again, we see the importance of the status of the present.

Those who have claimed that becoming is *subject*ive have argued that events are present, and therefore past and future, only in relation to a *subject qua* perceiver or language-user. I believe, and about this I am sweating with conviction, that events would undergo becoming whether or not they be perceived or spoken of. Rather than just turn in a wet sweat shirt, which is often the best we can do when we discuss ultimate issues in philosophy, I shall attempt to refute the arguments of those who believe otherwise. Both *a priori* and empirical arguments have been advanced, although not always with a careful distinction being made between the two. The *a priori* arguments are based on an analysis of the meaning of our concept of past, present and future, while the empirical arguments are based on consideration of contingent facts about non-linguistic features of the world. According to the *a priori* arguments it is a contradiction in terms to speak of events as being past, present and future in a world devoid of perceivers or language-users, while according to the empirical arguments it is only contingently false to speak in this way. In this chapter I shall try to refute the *a priori* arguments by arguing that our concept of an *A*-determination commits us to a realistic view of becoming: the following chapter will argue that this realistic commitment of ordinary language or common-sense could not be refuted by any empirical discovery.

A priori arguments have been advanced both for the mind-dependency and the linguocentricity of *A*-determinations. Bertrand Russell has claimed that these temporal notions have only a psychological meaning: when a person asserts that some event is past (present, future), he is not describing an objective property of this event but is merely *expressing* the particular species of belief he has towards the content of his belief, that is, whether it is a retrospective, perceptual or anticipatory belief.

> Difference of tense, in its psychologically earliest form, is not part of what is believed, but only of the way of believing it; the putting of the tense into the content is a result of later reflection. . . . Tense puts the time-relation, apparently, into the content of what is believed, whereas, if the above theory is correct, tense is primarily embodied in the nature of the belief-feeling.[1]

Tense, as well as other *A*-expressions, add nothing to the factual or informative content of an assertion: they serve only the pragmatic function of expressing the kind of belief the speaker has. I believe that an inspection of the general features of psychological terms, i.e. terms which refer to mental states or describe some feature of a mental event, can show us why *A*-expressions are not psychological terms.

In the first place, genetic considerations concerning how we come to understand the concept of *X* or, what amounts to the

[1] Russell, 'On Propositions: What They Are and How They Mean', in *Logic and Knowledge*, edited by R. C. Marsh (London, 1956), pp. 308–311. In his later work, Russell flirted with, although he did not completely embrace, behaviourism. He still held to the doctrine that *A*-expressions serve only to express the state of belief of the speaker—whether he remembers, perceives or anticipates what he reports—but he was more willing to give a behavioural analysis of these different species of belief: 'Thus the difference between a sentence beginning "this is" and one beginning "that was" lies not in their meaning, but in their causation. The two sentences "The Declaration of Independence was in 1776", uttered by us, and "The Declaration of Independence is in 1776", which might have been uttered by Jefferson, have exactly the same meaning, but the former implies that the causation is indirect, and the latter that it is direct, or as direct as possible.' *An Inquiry into Meaning and Truth* (New York, 1940), p. 113.

same thing, learn to use the word '*X*', are not relevant to determining whether or not '*X*' is a psychological term. In general, the meaning of a term cannot be determined by reference to how we learn to use it: for example, we learn numerical concepts by adding objects together, but it hardly follows that arithmetical concepts and operations refer to these empirical collections and the results attendant upon combining them together. I might learn to use the word 'pain' by reference to publicly observable occurrences, such as people grimacing and groaning, but it by no means follows that the word 'pain' is not used to refer to a private mental state or feeling. Russell once attempted to show that *A*-determinations are psychological notions by arguing that in order to understand them we must introspect our mental states: to understand the concepts of past, present and future we must refer respectively to one of our memories, sensations and anticipations.[1] Even if it were true, though it probably is not, that introspection was involved in our learning the concepts of past, present and future, it would not establish that these are psychological terms.

A word has a psychological meaning—describes a mental state or event—not in virtue of how it is learned but because the rules for its use display an asymmetry between first and third person. To explain what is meant by this asymmetry between first and third person discourse, let us consider the word 'pain', which unquestionably is a psychological term since it has the function of referring to a mental state or feeling. If someone says, 'He is in pain', we can ask him how he knows that the person he refers to is in pain; however, it is odd to ask someone who says, 'I am in pain', how he knows that he is in pain, for he simply has or experiences the pain and therefore does not need to infer it from evidence about overt bodily behaviour, as does a third person. The person who says, 'I am in pain', when he is neither lying nor misusing language, cannot be mistaken, unlike the person who says, 'He is in pain'. It is this asymmetry between first and third person talk which explains what is involved in the claim that we each have a privileged access to our own mental states and feelings,

[1] Russell, in his article, 'On the Experience of Time', *Monist*, 25 (1915), pp. 226, wrote: 'Thus introspection is necessary in order to understand the meaning of "past", because the only cases in which this relation is immediately given are cases in which one term is the subject. Thus "past", like "present", is a notion derived from psychology.'

If 'past', 'present' and 'future' have a psychological meaning they should display in their use this asymmetry between first and third person discourse. It is obvious that these terms display no such asymmetry and thus are not psychological. If I say that an event x is present other people do not say, 'Well, assuming that he is not lying or misusing language that settles it: he ought to know what he feels.' It makes sense to ask me how I know that x is now present, rather than past or future: and the methods I use for verifying that this is the case are the same as those which any other person could use. If I am standing in front of a window and remark that it is now raining, other observers can have exactly the same evidence for this as I can; all they have to do is come over to the window and look out. Hugo Bergmann has written:

> Even if one assumes . . . that the future is not uniquely determined by a temporal cross-section, one can say only that this indeterminacy prevails just as much for Plato as for myself and that I cannot decide by physical means who is living 'now'. For the difference is a psychological one.[1]

Bergmann's claim is fantastic, for it is not the psychologist but the coroner who determines whether it is I or Plato who exists now; and the techniques of a coroner are notoriously *physical.*

What this brings out is that whether or not an event (object) is now present (now exists) is subject to public verification, in which no observer enjoys a privileged status. But while no one observer enjoys a privileged status, there is a class of observers who enjoy such a status, namely those who presently exist. Only their testimony is relevant to our now determining the truth or falsity of my statement that it is now raining, and also for our now determining the objectivity or veridicality of my sense experience. Past and future observers will determine the truth of my present claim about the weather by reference to what they then observe, but this is irrelevant to how we *at present* determine what is the case. Our concept of the objectivity of sense experience involves the method by which we now test the veridicality of a sense experience, this test being based on an agreement in judgment

[1] Bergmann, *Der Kamp um das Kausalgesetz in der jüngsten Physik* (Braunschweig, 1929), pp. 27–28. Quoted from p. 323 of Adolf Grünbaum's *Philosophical Problems of Space and Time* (New York, 1963). In the next chapter we shall consider a way of interpreting Bergmann's comment which renders it not absurd.

between observers who exist *at present*. Since our concept of objectivity of experience presupposes that the observers whose testimony is relevant exist at present, it follows that our determination of presentness could not be subjective; for what test could establish that it is?

Some may have been led to classify *A*-expressions as psychological terms because of our use of the expression 'my past', 'my future', 'his past', etc. They may have thought the 'my' in 'my past' is that of mental ownership, as is the 'my' in 'my pain'. Just as there are no pains which are not some person's pain so too there is no past, for example, which is not someone's past. In the first place, this assimilation of 'my past' to that of mental ownership will not hold up. 'My past' refers to every particular thing which I have done and undergone; however, I did not have private access to all of these things, for many of them were publicly observable events, some of which I was not even conscious of, such as being born. The reason why my past cannot be someone else's past is that from the fact that I have done and undergone certain particular things it logically follows that no one else has done and undergone exactly the same things, although they may have done and undergone qualitatively similar things. In the second place, the concept of my past is narrower than that of the past; for there are things which are in the past but which are not part of my past, such as the death of Queen Anne. Many of my true statements about the past would become false if I affixed a 'my past' to them. There are past events, such as pre-historic ones, which are not part of any person's past. Therefore it is wrong to say that there is no past which is not someone's past. Similar considerations hold *mutatis mutandis* for 'my present' and 'my future'.

Not only are *A*-expressions not psychological terms, for the general reasons advanced above, but more specifically they are not used to express a species of belief or propositional attitude, as Russell claimed. To show that they are not, let us consider how their use differs from that of a word, such as 'believe', which does have the function of expressing our attitude towards some propositional content. Although the expression 'I believe' does, in some uses, have the illocutionary force of a humble operator which we affix to our statement to protect ourselves from having to answer the question how we know and to give ourselves an out if what we say is false, it nevertheless does describe our propositional attitude of acceptance towards the propositional content that is expressed by the noun 'that'-clause following it. When I say 'I believe that it is now raining' and we wish to verify whether or not I have truly described my state of mind towards the propositional content *that it is now raining*, we do so by observing whether my behaviour and disposition to behave are of the sort someone would display who had this psychological state of belief. We definitely do not verify this description of my psychological state of belief by observing the weather. But if someone says that some event is now past (present, future) we do not determine the truth of his assertion by reference to his behaviour and dispositions, unless the event he reports is one of his own mental states. It would be wrong to say that my statement that it rained is false because I am not in a psychological state of retrospectiveness or that my statement that it will rain in five hours is false because I am not in a state of anticipating an occurrence of rain.

It would seem, then, that Russell is wrong when he says that *A*-expressions are used to *express* the state of belief of the speaker. Rather it would seem that they describe or characterize the event which the speaker reports: they add some further truth-condition to the statement which is made. For instance, if I say 'Jones died' you could say that this is false, not because I am not accurately or truthfully describing my psychological state of belief—I might even have a distinct experience of seeming to remember this event—but because Jones is still alive: his death is not yet past. Thus the tense of the verb, which indicates in this case the pastness of the reported event, does add to the factual or informative content of my assertion; for the pastness of the event reported is one of the truth-conditions of the statement made. The informative function of *A*-expressions has already been explored at some length in Chapter IV.

In summary: the *a priori* argument for the subjectivity of becoming based on the claim that *A*-expressions are psychological terms fails because it is based on a misconstrual of the logic of *A*-expressions. Let us now consider a different *a priori* argument for the subjectivity of becoming, which is based on the claim that the subject is free to select or choose what event or time is present. This argument from the alleged selectivity of *A*-expressions is usually based on either the egocentric particular or token-reflexive analyses of *A*-expressions. First, we shall explain and contrast these two analyses.

Russell used the term 'egocentric particular' to refer to those words which have the peculiarity of picking out certain particulars from the universe by their relation to oneself.[1] Such words are tenses, pronouns and demonstratives. Russell attempted to analyse all these words in terms of the egocentric particular 'this', in which 'this' is a *logically proper name* for a sense datum experienced by the speaker at the time he makes his utterance. For example, 'I' means 'the biography to which this belongs'; 'now' means 'what is compresent with this'. 'This' is a strange sort of proper name because it applies to something different—a different sense datum—every time it is uttered. And yet on each occasion of its use it is unambiguous, applying to one and only one particular, though, of course, only the speaker can know its referent since it is a sense datum private to himself.

In accordance with Russell's account of egocentric particulars we could give the following analysis of *A*-statements, which shall be called the 'EP' analysis:

Analysandum	*Analysans*
X is now present.	*X is* simultaneous with this.
X is now past.	*X is* earlier than this.
X is now future.	*X is* later than this.

in which 'this' in the *analysans* is a logically proper name for a sense datum experienced by the speaker at the time he makes his utterance. Notice that what Russell's analysis does is to convert the seemingly monadic predicate 'is now present (past, future)' into a two-place relation in which one of the relata refers to a mental event—a sense datum. Also it is to be noted that 'this' functions as a temporal, rather than a spatial, indicator. The test for whether 'this' functions as a temporal indicator is whether it makes sense, and is natural, to say 'the time at which this occurs', as opposed to 'the place at which this occurs'. Since a temporal designation is being made by the above *A*-statements, 'this' functions in them as a temporal indicator, being short for 'the

[1] In 'The Philosophy of Logical Atomism', *Monist*, 28–29 (1918–19), Lecture IV, pp. 55–56, Russell refers to these words as 'emphatic particulars'. See chapter vii of *An Inquiry into Meaning and Truth* for Russell's most complete account of egocentric particulars.

time at which this occurs'; it would make no sense in these statements to interpret 'this' as meaning 'the place at which this occurs', since that would result in our saying '*X is* simultaneous with the place at which this occurs'. 'This' has the same logic as 'now' in these statements. On the other hand, in the statements '*X* is to the west of this', 'this' functions as a spatial indicator, it making sense to say '*X* is to the west of the place at which this occurs' but not 'the time at which this occurs'.

An independent, though in some ways similar, analysis of egocentric words was given by Reichenbach, who called them 'token-reflexive' words.[1] Reichenbach defines a word as a class of similar tokens, a token being a physical sign, such as ink marks on a piece of paper. By *similar* tokens is meant tokens that are equisignificant, rather than tokens that physically resemble each other. A token-reflexive word is one that refers to the corresponding token which occurs in an individual act of speech or writing. Every token of a given token-reflexive word will refer to a different physical token, namely itself. All such words can be defined in terms of 'this token'. For example, 'I' means 'the person who utters this token'; 'now' means 'the time at which this token is uttered or inscribed'. Every token of the phrase 'this token' refers to a different token—a different inscribing of a set of ink marks or uttering of noises. A sentence containing a token-reflexive word is called a token-reflexive sentence; every token of a given token-reflexive sentence will refer to a different sentence token.

In accordance with this treatment of token-reflexive words and sentences we can give the following analysis of *A*-statements, which has been referred to as the 'TRI' analysis:

Analysandum	*Analysans*
X is now present.	*X is* simultaneous with this token.
X is now past.	*X is* earlier than this token.
X is now future.	*X is* later than this token.

in which 'this token' refers to the uttering or inscribing of the entire sentence token within which the token of the phrase 'this token' occurs. It is important to realize that this TRI analysis is analysing a statement, i.e. what is said through the uttering or inscribing of the sentence which occurs in the *analysandum*.

[1] See Hans Reichenbach, *Elements of Symbolic Logic* (New York, 1947), pp. 284–287. The reader is referred back to the discussion of token-reflexivity on p. 18 ff.

According to this analysis, the statements made by simultaneously using the sentences in the *analysandum* and *analysans* are logically equivalent. 'This token', like the logically proper name 'this' in the *analysans* of the EP analysis, functions as a temporal, rather than a spatial, indicator because it is meaningful to say 'the time at which this token occurs' but not 'the place at which this token occurs'. It is meaningless to say '*X is* simultaneous with the place at which this token occurs'.

While the TRI analysis is in general agreement with the EP analysis, it differs from the latter in making the referent of 'this' the occurrence of the physical token, rather than some *other* event —a sense datum—which is simultaneous with the occurrence of this token. By doing this, the TRI makes explicit the implicit reference made to the occurrence of the sentence token in the *analysans* of the EP analysis, in that 'this' denotes a sense datum experienced by the speaker at the time he makes his *utterance*. Since 'this' is a logically proper name for an event occurring simultaneously with the occurrence of the sentence token containing 'this' we must first be able to locate the token of 'this' before we can know what its referent is. By making do only with the token of 'this' occurring in a speech act, the TRI analysis eliminates the need to bring in some event in addition to the occurrence of the token. Therefore this analysis is more economical than the EP one, which must make reference to two events—the occurrence of the token and a sense datum that is simultaneous with the token.

By now it is a familiar story that the statements in the *analysans* of the EP and TRI analyses are *A*-statements, even though they contain no tensed verb or copula. C. D. Broad, arguing from the erroneous assumption that an *A*-statement can be made only through the use of a tensed sentence, has claimed that the copula in the *analysans* is really in the present, rather than the timeless, present.[1] Broad argued that the 'is' in '*X* is simultaneous with this token' is in the present tense since it makes sense to say '*X* was earlier (preceded) this token'. I believe that it can be shown that the copula must be tenseless. In some contexts the copula 'is' is tenseless—for example, 'Two and two *is* four', 'Red *is* a colour'— while in others it is tensed—for example, 'The battle is raging'. The test for determining whether 'is' in some particular sentential

[1] Broad, *Examination of McTaggart's Philosophy*, *op. cit.*, p. 306.

context must function as a tenseless, rather than a tensed, copula is as follows: if the context permits the substitution of 'was' or 'will be' for 'is' the copula is not of necessity tenseless; if not it is of necessity tenseless. The qualification is made that the copula must be tenseless, or is of necessity tenseless, because there are sentential contexts in which the copula is tenseless, e.g. '*X is* simultaneous with *Y*', but in which it is not of necessity tenseless, for it makes sense to say '*X* was simultaneous with *Y*'. It is meaningless, for example, to substitute 'was' or 'will be' for 'is' in 'two and two *is* four', whereas such a substitution is not meaningless (though it may result in changing the truth-value of the statement) in 'the battle is raging'.

Using this test it can be shown that the copula must be tenseless in '*X is* simultaneous with (earlier than, later than) this token', for when 'was' or 'will be' is substituted for 'is' it results in the nonsensical utterance, '*X* was (will be) simultaneous with (earlier than, later than) this token'. But what exactly is wrong with this? We shall work with '*X* was (will be) simultaneous with this token', for the purpose of showing that it is contradictory. 'This' means 'what I am now pointing at'. Anything which I am now pointing at is, by definition, present. And anything which is now present is, by definition, not past or future. Anything simultaneous with something which is present is itself present, and therefore not past or future. Since *X* is simultaneous with what I am now pointing at, this event is present. But if it is now present it cannot be simultaneous with some past or future event, for in such a case it would itself be past or future, and therefore not present. Therefore, it is contradictory to say '*X* was (will be) simultaneous with this', for it entails that *X* is both present (since simultaneous with what I am now pointing at) and not present (since it was simultaneous with some event).

Having clarified what is involved in the EP and TRI analyses of *A*-statements, we can now consider the *a priori* argument for the subjectivity of becoming based on the alleged selectivity of *A*-expressions. This argument is *a priori* because it is based on an *analysis*—the EP or TRI analysis—of *A*-statements. Those who believe in the subjectivity of *A*-determinations deny that these determinations are intrinsic to events. All times are claimed to be equally real. There is 'no more a single rolling Now than there is a single rolling Here along a spatial line—a standing line of soldiers, for example, though each of them has his vivid presentment of his own Here'.[1]

Russell used his EP analysis of *A*-statements to justify this democratic way of conceiving time:

> I think it is extremely difficult, if you get rid of consciousness altogether, to explain what you mean by such a word as 'this', what it is that makes the absence of impartiality. You would say that in a purely physical world there would be a complete impartiality. All parts of time and all regions of space would seem equally emphatic. But what really happens is that we *pick out* certain facts, past and future and all that sort of thing; they all *radiate out from* 'this'.[2]

What is revealing in this quotation is Russell's use of the phrase 'pick out'. He claims that all parts of space and time are equally real, there being no 'here' or 'now' in the physical world. No event or moment of time is present intrinsically or simpliciter any more than any point of space is here simpliciter, but what happens is that the speaker 'picks out' or selects some event or moment of time by pointing to it and uttering the word 'this'. It would follow that what event or moment of time is denoted by 'this' is arbitrary, being subject to the speaker's free choice. Just as the spatial object denoted by 'this' is not more real than, does not exist in any different sense from, the objects that coexist with it in space, so too the event denoted by 'this' in the *analysans* of the EP analysis is not more real than, does not exist in any different sense from, the events preceding and succeeding it in time. Therefore, past, present and future are relative to the speaker, since he is free to choose what event will be denoted by 'this', and as a result qualify as being present. And since he is free to select the event which is to be present he also is free to select what is to be past and future. And for this reason *A*-determinations are subjective.

Ayer has given an analysis of *A*-expressions that involves the same 'picture':

> events are not in themselves either past, present, or future. In themselves they stand in relations of temporal precedence which do not vary with time. . . . What varies is only the point of reference which *is taken* to constitute the present . . . the point of

[1] D. Williams, 'The Myth of Passage', *op. cit.*, p. 467.

[2] Russell, 'The Philosophy of Logical Atomism', *op. cit.*, pp. 55–56. My italics.

reference, by which *we orient ourselves* in time, the point of reference which is implied by our use of tenses, is continuously shifted.[1]

What is revealing in this quotation is Ayer's use of the phrases 'the point of reference which *is taken*' and 'the point of reference, by which *we orient ourselves*'. Ayer, like Russell, suggests that the speaker chooses what event or moment of time is present; and since it is only in relation to this chosen event that other events are past, present and future, these *A*-determinations are subjective.

If the Russell–Ayer way of picturing the selectivity of 'this' is correct, it would follow that there is something subjective, in the sense of being speaker-dependent, about *A*-determinations. But this way of picturing the use or function of the demonstrative 'this' is incorrect and rests on a false analogy between a temporal series of events and a spatial order of objects. The crucial point is: the demonstrative 'this' is *spatially* selective, picking out one object in the speaker's visual field from among other objects coexistent with it which could have been chosen in its place. For example, in my visual field there are now three objects—a lamp, a pencil and a pen. I point to the lamp and say 'this', thereby picking it out; however, I could have chosen to point to the pencil or the pen and thus have picked it out instead of the lamp. But there is no temporal field of vision corresponding to a spatial field of vision, since events which are earlier and later than each other by definition do not coexist. While it is correct to say that 'now', as well as 'this' in the *analysans* of the EP and TRI analyses, function as temporal indicators or demonstratives, it would be wrong to say that they pick out, select or choose a moment of time.

What is so misleading in the quotation from Ayer is the suggestion that we are free to *orient ourselves* in time the way we do in space. The Russell–Ayer view of 'this' being temporally as well as spatially selective requires that we spatialize time by conceiving of events as quasi-substances that coexist and endure in a one-dimensional spatial order. The speaker on this view plays a role similar to the transcendent observer in the River of Time metaphor. He is outside this order and is free to 'move' along it in any direction he chooses, as is a general inspecting a row of soldiers. He can pick out any soldier he wants merely by walking up to him and saying 'this'. When Russell says that everything 'radiates out

[1] Ayer, *The Problem of Knowledge*, *op cit.*, pp. 152–153. My italics.

from "this"' he is treating 'this' as if it were the transcendent spotlight of presentness that plays along a line of coexistent substances, such as a row of soldiers. This River of Time metaphor, as has been seen, leads to absurdities, such as that time 'moves' at the rate of one time-unit per time-unit. The basic common-sense belief that man is not temporally free to choose his 'now', that he is finite and time-bound, is reflected in the fact that 'this' or 'now' is never temporally selective. We cannot 'move' about in time and make any event or moment of time we choose present by pointing to it and saying 'this'; we cannot alter our temporal perspective at will, as we can our spatial perspective. If we could choose our temporal perspective we could not have the concepts we do of planning, choosing, intending, causing something to happen, etc., all of which rest on the presupposition that there is an imposed temporal schema of past, present and future that is not of our doing or choosing. We cannot get out of problematic situations, which are the sort that call for deliberation and decision, by 'fleeing' to a different time.

A parallel argument to that of Russell–Ayer for the subjectivity of temporal perspectives, but based on the TRI rather than the EP analysis, would be the following: According to the TRI analysis '*X* is now present (past, future)' means '*X is* simultaneous with (earlier than, later than) this token'. This shows that events are past, present and future only in relation to a given utterance token, and since the speaker can choose or select which utterance token that shall be by his use of the token-reflexive phrase 'this token', it follows that *A*-determinations are subjective in the sense of being dependent upon the speaker.

This argument suffers from the same defective premiss as the Russell–Ayer argument—namely, that the speaker picks out or selects the event, in this case the occurrence of a sentence token, which will serve as the point of orientation for the temporal schema of past, present and future. When a person utters a word which by stipulation is token-reflexive, such as 'you' or 'this' in Reichenbach's analysis, he in no sense picks out or selects which particular token of 'you' or 'this' will be referred to. He has some choice, however, as to which person or object shall be denoted by his token of 'you' or 'this', assuming that there is more than one person or object in his immediate vicinity. 'This', when interpreted in accordance with the token-reflexive analysis, is a very queer sort of demonstrative, for it involves no element of selectivity. In the ordinary use of the demonstrative 'this' there is an element of spatial, but not temporal, selectivity. But if 'this' is token-reflexive, even the element of spatial selectivity drops out; for an utterance of 'this token', when interpreted as being token-reflexive, does not pick out one token from among other tokens simultaneous with it that could have been chosen in its place. What would it be like for a person to utter a token of the token-reflexive phrase 'this token' and not refer to the token he utters? It is impossible for a token of 'this token' to refer unsuccessfully, that is, not to refer to some physical token. It is possible for 'I' to be used unsuccessfully if uttered, for example, by a disembodied spirit, for it would merely prompt the question 'Who?' Similarly, a person who utters a token-reflexive sentence does not choose or pick out a sentence token.

In granting, for the sake of argument, that language could be token-reflexive, we may have conceded too much or, at any rate, may have overlooked the difference between the way in which a token-reflexive word, phrase or sentence refers to itself and the way in which it refers to something other than itself. I shall argue that: (i) no *word* is token-reflexive in the radical sense that no token of any word is self-referring; (ii) no *phrase* or *sentence* is token-reflexive in the less radical sense that for any phrase or sentence we can find some token of it which is not self-referring; and (iii) in cases where a particular phrase or sentence token is token-reflexive it does not refer to itself in the manner of picking out or selecting itself. In raising the question whether language can be token-reflexive we must stress that our concern is with ordinary language. One could make any word, phrase or sentence token-reflexive merely by stipulating that it shall be. We could stipulate, for example, as Reichenbach does, that any symbol is token-reflexive when written within what he calls 'token-quotes': ↘this↙. By writing 'this' within token-quotes we make it token-reflexive, but such a move is unilluminating and does nothing to show that the ordinary use of 'this' or any other symbol placed within token-quotes is token-reflexive.

(i) In arguing that no word is token-reflexive, we shall concentrate on demonstratives, pronouns and tensed verbs, since they are the usual candidates for token-reflexive words. The first thing to notice is that these words are used to refer not to themselves but to something other than themselves, though they do so in virtue of the spatio-temporal relation between the token occurring in a speech act and its referent. When I say 'you' I am referring to a person in my vicinity and not to the token of 'you' that I utter, though in order to know the referent of this token one would have to know the spatio-temporal relation that it has to the person in question. When I point and say 'this tree' no doubt I attract attention to myself and my utterance token; otherwise I could not refer successfully to the tree. But it would be wrong to say that I refer to myself or to the token I utter.

The demonstrative 'this' is ordinarily used in connection with an act of pointing, staring or some other bodily gesture. When used alone it is a surrogate for such an act of pointing, etc.: it is a stand-in for an index finger or the like. An index finger is a necessary element involved in pointing to something, but for this very reason can never point to itself, any more than a hammer, which is an instrument used in hammering something, can hammer itself. If I want to refer to my index finger I must point to it with my other index finger, or I might just hold up the index finger and say 'this index finger', but the 'this' is a surrogate for my pointing to this finger with my other finger. The only case in which an index finger might be said to point to itself is when a person uses his index finger to point to itself in a mirror. But then he points, not to his index finger, but to the image of his index finger in the mirror.

Since the demonstrative 'this' is a surrogate for an index finger, it follows that a token of 'this' is never self-referring any more than an index finger is. When 'this' is used in conjunction with some other word(s)—for example, 'this tree', 'this token'—'this' can never refer to the token of 'this' which occurs in the speech act, but always to something other than the token. If a perverse radio announcer should say, 'The time will be five o'clock when you hear the next token of the word "this". . . . This', it would not be correct to say that the token of 'this' occurring at the end of his sentence token is self-referring, for he denotes or refers to this particular token by the definite description, 'the next token of the word "this",' that immediately precedes it. Another possible counter-example would be to say 'this room and everything in it', but here 'this' refers not to itself but to a room, it not being essential that the token of 'this' occur in the room referred. It would not if the speaker were standing outside the room when he made this utterance. Similar considerations can be brought to bear against the counter-example 'this phrase and every token in it'; for 'this' refers to something other than itself—a phrase—and, moreover, it is not essential that the token of 'this' occur within the phrase referred to.

(ii) So far we have argued that in ordinary language no *word* is token-reflexive in the radical sense that no token of any given word is self-referring. The token-reflexive theory does not, however, claim merely that certain words, when used alone, are token-reflexive, but also that their inclusion in a *phrase* or *sentence* renders that phrase or sentence *as a whole* token-reflexive. In attempting to answer the question of whether a phrase or sentence as a whole can be token-reflexive, we must repeat the definition of a token-reflexive expression which was given on pages 197–198: a token-reflexive word, phrase or sentence is one that refers to the corresponding token occurring in an individual act of speech or writing. Therefore, *every* token of a given token-reflexive word, phrase or sentence must refer to itself. I believe that it can be shown that no phrase or sentence is token-reflexive in this sense, because for any phrase or sentence we can find some token of it which is not token-reflexive. In this connection let us consider the sentences which are most often cited as being token-reflexive: 'This sentence is in pica type', 'This sentence contains five words', 'This sentence is short', 'This sentence token is loud'. Most tokens of these sentences are not self-referring because they are used to refer to a sentence token which immediately precedes or follows them; for example, 'This sentence contains five words. "The boy hit the ball".' Tokens of these sentences are self-referring only in default of there being any other sentence token in their immediate vicinity for them to refer to; for example, if a token of the sentence 'This sentence is in pica type' appears alone on a sheet of paper we then assume that it is self-referring. Thus token-reflexivity is a kind of degenerate case of referring. There are some sentence tokens, such as 'This sentence is false', which we would not interpret as being self-referring even in default of there being some other sentence token for them to refer to. Similar considerations apply to phrases such as 'this phrase'.

(iii) In cases where a particular phrase-token or sentence-token is token-reflexive we must make two things clear. First, a token-reflexive phrase or sentence token refers to itself *qua* physical token and not to itself *qua* semantic token, it being nonsense to say 'The meaning of this sentence is the meaning of this sentence' or 'This sentence refers to what this sentence refers to'. Secondly, the word 'refer' is being used in a different way from when we say that a demonstrative, such as 'this', or a pronoun, such as 'you', refers to something; for only in the latter case is there an element of selectivity—the picking out of one thing from among other things which could have been denoted in its place. Since the argument for the subjectivity of *A*-determinations rests on confounding these two different senses of 'refer', let us designate the latter as 'refer$_1$' and the former as 'refer$_2$'. A token-reflexive phrase or sentence token refers$_2$ to itself, but cannot refer$_1$ to itself because it does not pick out or select itself. In cases where it has been stipulated that 'this' or 'this token' is to be token-reflexive, we must be careful not to confuse the way in which such a word refers to the token occurring in an individual speech act and the way in which it refers to something other than the token occurring in a speech act. A token of a token-reflexive word refers$_2$ to itself and refers$_1$ to something other than itself. But unless it can be shown that an *A*-sentence token refers$_1$ to itself it will not have been shown that *A*-determinations are subjective in the sense of being subject to the selection of the speaker.

A defender of the *B*-Theory's Psychological or Linguistic Reduction might agree that our use of *A*-expressions involves no element of selectivity but still argue that the EP or TRI analysis can be appealed to in support of the *a priori* claim that becoming is subjective. The EP analysis of '*X* is now present (past, future)' into '*X is* simultaneous with (earlier than, later than) this sense datum' shows that events have *A*-determinations only in relation to a mental event, while the TRI analysis of the former statement into '*X is* simultaneous with (earlier than, later than) this token' shows that events have *A*-determinations only in relation to a linguistic event. If there were no subjects or persons in the world there would be no mental or linguistic events, and therefore there would be no becoming. I shall try to refute this new *a priori* argument by showing that the EP and TRI analyses fail to give an adequate explication of our ordinary concept of an *A*-determination.

I shall assume that both the EP and TRI analyses purport to

analyse a statement into another statement that is logically equivalent to the *analysandum*. According to the EP analysis that *X* is now present is logically equivalent to that *X is* simultaneous with this sense datum. I believe that it can be shown that the *analysans* of this analysis is not logically equivalent to the *analysandum*. If two statements are logically equivalent they must have the same entailment relations: any statement which entails one of them entails the other and any statement entailed by one of them is entailed by the other. However, *prima facie* it seems that the statement that *X* is now present and the statement that *X* is simultaneous with this sense datum do not have the same entailment relations: the latter, unlike the former, entails that there is a sense datum, and, moreover, one which is simultaneous with *X*. The statement that *X* is now present although not simultaneous with any sense datum does not appear to be a contradiction; but according to the EP analysis it is, for it is analysed into the statement that *X is* simultaneous with this sense datum although not simultaneous with any sense datum. It would seem, therefore, that our ordinary concept of being present does not involve being perceived or being simultaneous with a sense datum. In this matter the commitments of common-sense or ordinary language are realistic.

The same sort of objection can be raised against the TRI analysis; for that *X* is now present, unlike the statement that *X is* simultaneous with this token, does not entail that there is a token, and moreover one which is simultaneous with *X*. The statement that *X* is now present seems to have different truth-conditions from the statement that *X is* simultaneous with this token; for that a token occur simultaneously with *X* is a truth-condition of the latter but not the former statement. The statement that *X* is present although no token occurs, unlike the statement that *X is* simultaneous with this token although no token occurs, does not appear to be a contradiction. The former statement, unlike the latter, could be true. What is peculiar about the statement that *X* is now present although no token occurs is that while it is a contingent statement—it could be true—no one could truly assert it; for in the very act of asserting it a token is produced and the statement thereby falsified. The sentence which is used in making this statement might appropriately be called a 'pragmatically self-falsifying sentence', since any use of it will make a false statement, though, to be sure, a contingently false statement. Another famous example of a pragmatically self-falsifying sentence is 'I do not exist'. My existence is not necessary: it could be true that I do not exist, only I could never truly assert that it is so.

In a declarative speech act it is crucial to distinguish *what is said*—the statement made—which I shall call the *semantic* content of the declarative speech act, from the conditions which must be fulfilled both for the speaker to truly assert this statement and for the hearer to understand what statement is being made. These conditions shall be called *pragmatic* conditions of the declarative speech act. The thesis which I shall defend is that *A*-expressions are semantically objective but pragmatically subjective. What this means more precisely is that the statement made—the semantic content—in a declarative speech act involving the use of an *A*-sentence does not refer to a token or even entail the existence of a token; however, on the pragmatic level a *special* sort of reference must be made to the utterance token which occurs in this speech act.

As we saw above, the statement made in a declarative speech act involving the use of the *A*-sentence, '*X* is now present', which is the statement that *X* is now present, does not refer to or entail the existence of a token. But the pragmatic conditions which must be fulfilled for a speaker truly to assert this statement and for a hearer to understand what statement is being made include a special sort of reference to the utterance token produced in the speech act. Let us see why, with attention being paid first to the pragmatic conditions for truly asserting an *A*-statement.

In order to make a true statement, regardless of what subject matter it is about, it is trivially true that a person must say or write something.[1] To make the statement that *X* is now present, as well as the statement that two and two *is* four, a person must token some sentence. However, in regard to the pragmatic conditions for truly asserting an *A*-statement there is a special sort of reference to the sentence token produced in the speech act which does not arise in regard to asserting truly that two and two *is* four. This special sort of reference to a sentence token is brought out by the Truth-condition Rules for *A*-sentences, which were developed in Chapter III. These Rules specify that a present (past, future) tense sentence can be used to make a true statement only if tokened simultaneously with (later than, earlier than) the event

[1] We shall count the mental tokening of a sentence as making a statement. For ease of exposition reference to this special case of stating will be omitted.

reported by the statement made. These Truth-condition Rules for *A*-sentences, unlike those for sentences like 'Two and two *is* four', specify that a specific sort of *B*-relation must hold between the utterance token produced in a speech act and the event reported if a true statement is to be made. This is why we have said that the pragmatic conditions for truly asserting an *A*-statement contain a *special* sort of reference to an utterance token.

It must be emphasized that the Truth-condition Rules for *A*-sentences, while specifying the conditions under which *A*-sentences can be used to make a true statement, do not give the truth-conditions for the *A*-statements which are made through the use of these sentences. As insisted upon above, the statement that *X* is now present does not have as one of its truth-conditions that an utterance token occur simultaneously with *X*: it could be true that *X* is now present even though no one reports this fact. It is vital, therefore, to distinguish between the truth-condition rules for *A*-sentences and the truth-conditions of the *A*-statements that are made through the use of these sentences. The TRI analysis confounds these two different type of truth-conditions: it erroneously tries to pack the pragmatic conditions which must be fulfilled for truly asserting an *A*-statement into the statement which is made. What is curious about the EP and TRI analyses is that the sentences in the *analysandum* and *analysans* obey the same truth-condition rules, and yet the statements made by simultaneously tokening these sentences are not logically equivalent; that is, the sentences '*X* is present' and '*X is* simultaneous with this token (this sense datum)' can be used to make a true statement only if tokened simultaneously with the occurrence of *X*, but do not make logically equivalent statements.

The pragmatic conditions which must be fulfilled for a hearer to understand what *A*-statement is being made by a speaker also contain a special sort of reference to an utterance token. Once again it is trivially true that in order to understand any statement which a speaker makes, we must refer to the sentence token he produces: we must examine his utterance token in order to determine what words it contains and how they are connected together. What is special about the addressee's reference to an utterance token when determining what *A*-statement a speaker has made is that only in this case must he know when in time the speaker's utterance token occurs in relation to his hearing or seeing it. For example, if I see or hear an *A*-sentence token, such as 'The market is now falling', I would know what *A*-statement was made by the person who uttered or inscribed this sentence token only if I know when this token was produced in relation to my seeing or hearing it, that is I must know that it was produced virtually simultaneously with my seeing it or I must know that it was produced so many time-units earlier than my seeing it.[1] We are all familiar with the problem of understanding what *A*-statement was made by someone who tokened an *A*-sentence that we see in print or hear on a tape-recorder.

I hope that this discussion brings out the sense in which *A*-expressions are semantically objective and pragmatically subjective. The token-reflexive theory confounds the semantic and pragmatic dimensions of a declarative speech act. A similar sort of confusion underlies one of Berkeley's *a priori* arguments against the possibility of a physical object existing unperceived and unthought of. Berkeley argued that it was a manifest contradiction to conceive of a tree existing now unperceived and unthought of because you are thinking of it all the while. Plainly this is to confuse the semantic content of the conception (the what-is-being-conceived), which is *a tree existing now unperceived and unthought of*, with the pragmatic conditions that must be fulfilled for someone to entertain this conception, which trivially involves that he thinks something. A contradiction would result only if one conceived of *a tree existing now unperceived and unthought of and thought of by me*. Berkeley's argument, when translated from the idiom of conceiving into that of stating, takes the form that it is a manifest contradiction to assert that a tree exists now unperceived and unthought of since you are thinking of it all the while. The statement *that a tree exists now unperceived and unthought of* does not seem to be a contradiction: this statement could be true only it could never be truly asserted since the sentence used in making this statement is pragmatically self-falsifying.[2] The same point goes through in

[1] This assumes that the person who produced this sentence token was making a statement, rather than making-a-statement-to-be. See p. 154.

[2] This holds only if we include intentional elements in the *use* or *assertion* of a sentence, i.e. that the person who tokens the sentence is conscious of what he is saying. A distinction is to be made between a use (in this intentional sense) and a mere physical tokening of a sentence. Notice that 'I do not exist' is not pragmatically self-falsifying if by a pragmatically self-falsifying sentence we mean a sentence every *physical tokening* of which makes a false statement.

regard to the statement *that a tree exists now unperceived, unthought of and unspoken of*. 'To be present' no more means to be spoken of than it does to be perceived. Whereas Berkeley had to bring in the faithful perceiver God to save his system from absurdity, I fear that a defender of the TRI analysis would have to postulate a chatter-box God, who incessantly reported what was happening. This is the conception of God as the Great Sportscaster in the Sky. I conclude that any *a priori* argument which attempts to show that it is logically impossible for events to be past, present and future in a world devoid of perceivers or language-users is wrong, for it is based on an incorrect analysis of the meaning of *A*-expressions.

It might be objected that my distinction between the semantic and pragmatic elements of a declarative speech act commits me to the acceptance of the very mooted realistic theory of propositions or statements according to which propositions or statements, as the what-is-said in a declarative speech act, are denizens of some non-spatio-temporal 'third realm' in which they subsist independently of language and thought. According to this theory there are an infinite number of such timeless statements, both true and false, since there are an infinite number of things that could be stated or said in a declarative speech act. This Third Realm turns out to be an overcrowded slum, which upsets those metaphysicians with a love for desert landscapes, as well as metaphysicians possessed with 'a vivid sense of reality'. What have I said that might commit me to this view of propositions? It will be recalled that I said that it could be true *that X is now present even though no token occurs*, only no one ever could truly assert this statement. Does this not commit me to the view that statements are abstract entities whose being is independent of their being asserted, for it appears that they could be true even if never asserted.

Two things should be said in answer to this objection. First, IT IS TRUE *that it could be true that X is now present even though no token occurs*. And *if* this statement entails the realistic theory of statements, then this theory is the correct one, regardless of our esthetic predilections for an unbloated world. However, I do not think that this statement does entail the realistic theory. To show that it does not note that this statement is compatible with a nominalistic theory of statements, according to which a statement is a 'logical construction' out of sentence tokens. On the nominalistic account *that there is a statement* entails *that there is a sentence token. That X is now present even though no token occurs* does not entail, even on this nominalistic account, *that there is a token*; for it does not entail *that there is a statement. That X is now present even though there are no statements* is not, even for a nominalist, contradictory, although, on his theory, it cannot be truly stated for it involves the use of a pragmatically self-falsifying sentence. The statement *that there is the statement that X is now present even though no token occurs* entails, on a nominalistic account, *that there is a token*, and therefore the statement is contradictory. Therefore, even for a nominalist, it could be true (i.e. it is logically possible) *that X is now present even though no token occurs*; therefore this statement does not entail the realistic theory of statements.

A defender of the TRI analysis might claim that the statement that *X* is now present even though no token occurs really is contradictory; we cannot always trust our linguistic intuitions, since some contradictions are not obvious. He might give this argument. The first premiss is the unexceptionable statement that if an event is past (future) than some event is present. It then uses as its second premiss our reduction of *B*-relations to *A*-determinations from page 92:

(II) *P is* earlier than *Q* . ≡ . *P* is past and *Q* is present *or P* is past and *Q* is future *or P* is present and *Q* is future *or P* is more past than *Q or Q* is more future than *P*

Disjuncts two, four and five of this exclusive disjunction contain events that are past and/or future but no event which is present. If *P* and *Q* were the only two events in history and one of these three disjuncts were to be the case it would violate our initial premiss that if an event is past (future) some event is present. To ensure against this happening we must bring in a reference to the utterance token which occurs in the making of the above disjunction: for example, disjunct five *that Q is more future than P* could be analysed as *that P* is *earlier than Q and P* is *later than this token*. This is an interesting argument but does nothing to show why we must accept the TRI analysis; for all we have to do is to say that if disjunction two, four or five is the case, then it is not the case that *P* and *Q* are the only two events in history. The solution employing the TRI analysis also makes this move, for it brings in an event—the occurrence of the token—in addition to P and Q. If disjunct two, four or five is true than there is a present event, but it need not be a linguistic event, nor for that matter a mental event.

Before trying to refute the empirical arguments for the subjectivity of becoming, I would like to consider the following sort of objection to my claim that there is an objective present or now in nature. This objection, which I am sure the reader has wanted to raise at several places in this book, assumes that there is no objective here in nature. I grant this assumption. This objection then attempts a *reductio ad absurdum* of my claim by showing that all of the considerations I have advanced for the objectivity of *A*-determinations also hold true for here. So far in this book I have shown that: (*a*) *A*-expressions, although not sensible terms, still have an informative function, and for this reason a statement made through the use of a sentence containing a temporal indexical term (i.e. an *A*-expression) cannot be rendered without loss of meaning by a statement made through the use of a sentence containing no temporal indexical sign; (*b*) *A*-expressions are not selective; and (*c*) *A*-determinations are semantically objective but pragmatically subjective.[1] But all this holds true for here: (*a′*) 'here', although not used to refer to a sensible property, does have an informative role, and for this reason a statement made through the use of a sentence containing a spatial indexical sign cannot be rendered without loss of meaning by a statement made through the use of a sentence containing no spatial indexical signs; (*b′*) 'here' is no more selective than is 'now', in that just as 'now' cannot help but indicate the time at which the speaker's utterance occurs the use of 'here' cannot help but indicate the place occupied by the speaker; and (*c′*) 'here' and 'there' are semantically objective but pragmatically subjective, since that X is here does not entail that there is either a speaker or a token but to assert truly that this is the case there must obtain a specific spatial relation between X and the speaker, just as to truly assert that X has a certain *A*-determination there must obtain a specific temporal relation between X and the occurrence of the speaker's token.

While points (*a′*) and (*c′*) are correct, this objection falters in

[1] It has already been shown on p. 77 that here, like now, cannot be ostensively defined. 'Here' and 'now' are neither descriptions nor names, though they do function respectively as spatial and temporal indicators.

(*b*′), and also overlooks several significant disanalogies between temporal and spatial indexical signs, which shall now be mentioned. There are at least two senses in which it is not true that 'here' is no more selective than 'now'. First, while it is true that an utterance of 'now' cannot help but indicate a time which is simultaneous with or contains the occurrence of the token of 'now', it is not true that an utterance of 'here' cannot help but indicate a place which coincides with or contains the place occupied by the speaker.[1] 'Here', when used in conjunction with an act of pointing, is just as spatially selective as is the use of 'this'. For example, I say to the janitor, 'Put the chair here', as I point to a place several feet from my body. Of course, when 'here' is not used in conjunction with an act of ostension, it is not spatially selective. But even with non-demonstrative uses of 'here' there is a sense in which it is selective and 'now' is not. By moving with a certain velocity I can, within certain physical limits, choose the place denoted by my use of 'here' *n* time-units from now. If my velocity is zero I remain where I am and as a result denote the same place by my subsequent use of 'here'. The analogue for the selectivity of 'now' would have to be: By moving with a certain velocity I can, within certain physical limits, choose the time denoted by my use of 'now' *n* space-units from here. If I choose to move to this spot quickly I shall denote an earlier time by my use of 'now' than if I move to this spot slowly; however, I must move with a finite velocity or I shall never get to this other place. Herein is the disanalogy, since in the former case I could choose to have a zero velocity and just stay put.

My next utterance of 'now' will denote a different time, even if I just wait where I am, but my next utterance of 'here' will not denote a different place unless I move about. This difference between here and present or now is due to the fact that there is no spatial analogue to temporal becoming: the present (now), unlike here, shifts inexorably, independently of what we do. Every event later than the present will become present and every event earlier than the present did become present, to which the

[1] This claim is qualified in both cases by 'contains' because 'now' and 'here' are accordian terms, the width or extent of the time and place they respectively denote varying with the context. In some contexts 'now' denotes a time which overlaps the occurrence of the token, while 'here' often refers to a place which contains the speaker, e.g. 'Having a wonderful time. Wish you were here.'

spatial analogue would be that every object in front (to the right, etc.) of me will occupy (become) here and every object in the rear (to the left, etc.) of me will occupy here. But whereas the former is necessarily true the latter is contingent, and what is more is almost certainly false.

In addition to our spatial perspective being selective in senses in which our temporal perspective is not, it is also presupposed that we share the same temporal but not the same spatial perspective. It was seen that our test for the objectivity of our present sense experience, which is based on the agreement in judgment between different observers, presupposes that these observers exist at present and therefore share the same temporal perspective; but they need not, and usually do not, share the same spatial perspective, that is they are not all standing at the same place. The observers whose testimony is relevant to determining the truth of my claim to now be seeing a cat on the mat must exist now, but they need not be standing here. Also our use of tensed or *A*-questions—questions that require an *A*-statement for their answer—presupposes that the communicants share the same temporal but not the same spatial perspective. This is brought out by the fact that it is meaningful to ask 'Where are you?', to which the answer could be 'I am here': herein it is the direction that the voice comes from that constitutes the answer. However, there is no use for the question 'When are you?': 'I am now' has no use in our language. The reason why it is meaningless to ask this question is that it is a precondition for the asking of an *A*-question that the communicants share the same present, although not the same here, and obviously this precondition cannot itself be the subject of an *A*-question, as it is when we ask someone 'When are you?'

There is another difference between spatial and temporal indexical signs that further reinforces my claim that now or the present is objective in ways in which here is not. Let us consider the sentences '*X* is here' and '*X* is there', which are differently spatially indexed counterparts of each other. It is possible for these sentences to be used so as to make the same statement in a tougher sense than logically equivalent statements. For example, *A* says '*X* is here'[1] and at the same time *B* says, as he points to the spot where *A* is standing, '*X* is there'. Certainly *A* and *B* have made the same statement, and if we asked either of them what statement

[1] Assume that this is a non-demonstrative use of 'here'.

they had both made he would be willing to indicate what statement that is by making use of the appropriately spatially indexed counterpart sentence: *A* would say, 'We both said that *X* is here', while *B* would say, 'We both said that *X* is there'. Consider, on the other hand, the differently temporally indexed counterpart sentences '*X* is present today' and '*X* is future (past) by one day'. Can these sentences ever be used to make the same statement in an equally tough sense of 'make the same statement' as in the preceding case? Suppose that *A* today says '*X* is present today' and that yesterday (tomorrow) *B* says '*X* is future (past) by one day'. Certainly they have made logically equivalent statements: a statement endorsing either of these statements, as seen on pages 140–141, entails the other. But they have failed to make the same statement in the tougher sense.[1] For neither *A* nor *B* would be willing to answer the question, 'What statement did you both make?', by using the appropriately temporally indexed counterpart sentence; that is, *A*, for example, would not reply to this question by saying 'We both said that *X* is present today', nor would he reply 'We both said that *X* would be present today', because this makes it look as if both he and *B* made a prediction in the past.

[1] Since *A* and *B* have made logically equivalent statements, *A*-statements are not so hopelessly 'fugitive' that we cannot agree with and dispute the *A*-statements made in the past (and future). See A. Duncan-Jones, 'Fugitive Propositions', in *Philosophy and Analysis*, edited by M. Macdonald (New York, 1954).

XI

EMPIRICAL ARGUMENTS FOR THE SUBJECTIVITY OF BECOMING

IN my discussion of the *a priori* arguments for the subjectivity of becoming I tried to show that both the EP and TRI analyses misconstrue the logic or meaning of ordinary temporal discourse, which, I claimed, commits us to a realistic view of becoming. Whether or not this common-sense realism is naïve, in that it cannot be squared with the *empirical* findings of modern science, is completely irrelevant to what is ordinarily meant by *A*-expressions.

It might be thought that a more promising way of showing that becoming is subjective is to advance *empirical* arguments based on scientific discoveries. It has often been claimed that modern physics reveals the mind-dependency of *A*-determinations, and therefore the subjectivity of becoming. No similar claim has been made for the linguocentricity of becoming: the reason why will become obvious as the discussion progresses. Often the authority of the theory of relativity is appealed to in support of the claim that becoming is psychological or mind-dependent, as it is by the following authors:

> The objective world simply *is*. It does not happen. Only to the gaze of my consciousness crawling upward along the life (world-) line of my body does a section of this world come to life as a fleeting image.[1]

[1] H. Weyl, *Philosophy of Mathematics and Natural Science* (Princeton, 1949), p. 166.

> Events do not happen; they are just there, and we come across them. 'The formality of taking place' is merely the indication that the observer has on his voyage of exploration passed into the absolute future of the event in question; and it has no important significance.[1]

> Instead of allowing for the transient division of time into the past and future by the shifting Now of experienced time, the theory of relativity conceives of events as simply being and sustaining relations of earlier and later, but not as 'coming *into* being'; we conscious organisms then 'come across' them by 'entering' into their absolute future, as it were. And upon experiencing their immediate effects, we regard them as 'taking place' or 'coming into being'.[2]

There are two questions which ought to be raised concerning these psychological representations of becoming. First, exactly what empirical features of the world or what features of physical theory require us to classify becoming as mind-dependent? And, second, how adequate are these psychological representations of becoming? In answer to the first question it will be contended that there are no empirical facts which support the thesis that becoming is mind-dependent and that, appearances to the contrary, only *a priori* arguments, which are unsound, have been given for this thesis. The answer to the second question will try to show that there are serious conceptual difficulties in these psychological representations of becoming.

Before attempting to deal with the first question concerning what empirical facts or features of physical theory could support the claim that becoming is psychological it should be pointed out that it is not scientific theories which *conceive* of the *objective world* as being a certain sort of way but philosophers who do, though, to be sure, they may utilize such theories in support of their way of conceiving of the world. It is not the theory of relativity which 'conceives of events as simply being. . . ,' as Grünbaum claims, but rather Grünbaum, the philosopher, who conceives of the world as being thus. The crucial distinction between a scientific theory and the philosophical interpretations of its significance is

[1] A. S. Eddington, *Space, Time, and Gravitation* (Cambridge, 1953), p. 51. As we shall see, Eddington attacked this view because he believed that becoming had a physical basis in the indeterminism of modern physics. What is of significance here is that he thinks that this view is supported by a deterministic physics.

[2] A. Grünbaum, *Philosophical Problems of Space and Time*, *op. cit.*, pp. 318–319.

blurred by the above authors. Also, it is difficult to see how a physical theory, such as relativity, which is concerned with spatio-temporal measurements, could inform us as to the manner in which the mind or consciousness is related to the physical world. Plainly, these psychological representations are not part of the theory of relativity, but are philosophical interpretations of its metaphysical significance.

It might be thought that the mind-dependency of *A*-determinations is a result of the empirical fact that signals, such as light rays and sound waves, have a finite velocity. It is only a *contingent* fact that our perception of events outside our brain is not simultaneous with these events, for the opposite of this is certainly conceivable. Because of this contingent fact the events which two different observers perceive as being co-present need not coincide. The statement that two spatially separated events are absolutely simultaneous, i.e. simultaneous without any reference to a particular frame of reference, unlike the statement that everything in the universe (including measuring rods) doubled in size, is not absolutely meaningless; it only lacks operational meaning in our world with its signals of finite velocity. In a world in which there were no temporal lapse between our perception of an event and the occurrence of the event it would have operational meaning as well. This distinction between absolutely meaningless assertions and assertions which lack operational meaning because of some contingent feature of the world may account for why it is that the ordinary person does not think it nonsensical to assert that two spatially separated events are absolutely simultaneous.[1]

[1] The only attempt I know of to use the fact that signals have a finite velocity to support the claim that 'now' has a psychological meaning is in an article by a medical doctor, Robert Efron, entitled 'The Duration of the Present', in *Proceedings of the New York Academy of Sciences Conference on Interdisciplinary Perspectives on Time* (New York, 1967). On page 713 he writes: 'A star, two million light years distant from the earth, which exploded two million years ago, is experienced as exploding "now" by an observer on earth. Similarly, an electric shock, delivered to our great toe and experienced by us "now", was actually given at least 25 milli-seconds previously. From these two examples of the use of the word "now", we see that it refers to the first moment of awareness of an event or an entity by a particular observer who is located in a unique position in the universe. "Now" is a term which refers, therefore, to an aspect of the *experience* of existence. This aspect of our experience is implicit in the very state of awareness and in every state of awareness. To be conscious *is* to be aware of something *now*. To be aware of nothing *now*, is to be unconscious.' The surprising thing about this is that Efron is trying to find out what we mean by the word 'now' by reference to facts about the physics and neurology of perception rather than facts about linguistic usage! A dictionary turns out to be a work in

There is no question that the contingent fact of the relativity of simultaneity of spatially separated events and the relativity of precedence between events which are not causally connectible raises many important problems for ordinary temporal discourse, which very likely will require a reconstruction of certain areas of ordinary discourse. For example, in ordinary language 'earlier than' and 'simultaneous with' are dyadic relations, no mention having to be made of a third position from which one event is earlier than or simultaneous with another. But due to the finite velocity of signals there will be occasions on which our use of these temporal expressions will be ambiguous. To eliminate this ambiguity reference will have to be made to a frame of reference, and thus these expressions will now express a triadic relation. The expressions 'see', 'hear', etc., will also have to be refined when applied to perception of distant events. However, our concern here is only with the question of how this relativity of simultaneity resulting from the finite velocity of signals supports the claim that temporal becoming is mind-dependent. It is significant that those philosophers of science who claimed that physics reveals the mind-dependency of becoming have *not* supported their claim by reference to the contingent fact that signals possess a finite velocity. No doubt, they have not done so because it is obvious that being relative is not the same as being subjective, as Wilfrid Sellars has said. Temporal relations, also would have to be classified as subjective by this argument, for they also display in certain cases a relativity to a frame of reference. If someone were so foolish as to claim that becoming is mind-dependent because of the contingent fact that signals have a finite velocity he would

---

natural science. There is no question but that Efron has not articulated the ordinary meaning of 'now'; for it is not contradictory to say that the supernova which we are now experiencing is not occurring now but what is happening now is that there is a diffuse mass of gas surrounding the place where we see the supernova. In fact, it is not even contradictory to say that an event is occurring now though not perceived at all. The last two sentences in this quotation are correct. But because the concept of now or presentness is contained in our concept of consciousness does not establish that the converse is the case. The most charitable way of construing Efron's remarks is that he is seeking to replace our ordinary meaning of 'now' with one which will be scientifically respectable. But if this is what he is doing he fails to give any reason for why 'now' must refer to an act of conscious awareness because of the contingent fact that our awareness of events is not simultaneous with these events. If he can give no reasons for his claim that 'now' refers to conscious awareness he has no right to say this: philosophy is a put-up-or-shut-up activity.

have to be prepared to admit that if, as is logically possible, there were to be no time lapse between our perception of an event and the occurrence of the event becoming would not be mind-dependent. And this no defender of the mind-dependency thesis has been willing to admit.

Also irrelevant to the question of the objectivity of becoming is the question of whether time is discrete or mathematically continuous. Our belief in the objectivity of becoming does not require us to deny the continuity of time, nor require us to define a punctal event in terms of events of finite duration. It might be argued that the temporal continuity of time is not consistent with the objectivity of becoming because it is an essential part of the meaning of the ordinary concept of becoming that time is discrete, so that it would be a contradiction in terms to speak of a continuous temporal becoming. But certainly no one would want to advance the exactly parallel argument that since an essential part of the ordinary meaning of 'earlier than' is that time is discrete, each event having an immediate successor, it is meaningless to speak of a punctal event as being earlier than another punctal event in a continuous time. If it is true that our ordinary concept of time involves the discreteness of time, though it is by no means clear that it does, then if we are to speak meaningfully of becoming as continuous we must drop this feature of our ordinary concept of time. However, we can retain all the remaining features of this ordinary concept, for example that the present is objective, just as, analogously, we retain all of the features of the ordinary concept of *earlier than* when we speak of punctal events ordered by the relation of *earlier than* in a continuous time except that it is no longer required that each event have an immediate predecessor.

As far as I can make out, the defenders of the mind-dependency thesis have advanced only *a priori* arguments, although at times this does not appear to be the case. On some occasions they support their thesis by recourse to arguments which appear to be *a priori* since they are based on an analysis of the meaning of the concept of presentness. An example of this is Hugo Bergmann's claim that the concept of 'I' 'is inextricably intertwined with the concept "now". . . ."Now" is the temporal mode of the experiencing ego.'[1] This seems to amount to nothing more than the

[1] H. Bergmann, *Der Kampf um das Kausalgesetz in der jüngsten Physik*, *op. cit.*, quoted from p. 323 of Grünbaum's *Philosophical Problems of Space and Time*, *op. cit.*

a priori claim that the concept of presentness or Now involves a reference to a conscious subject as part of its very meaning; which claim has already been discredited as resting on an inadequate analysis of ordinary language. In developing Bergmann's position, Grünbaum wrote that 'the "flux of time" or transiency of the "now" has a *meaning* only in the context of the egocentric perspectives of sentient organisms'.[1] And further on he says that 'this instantaneous awareness of succession . . . is an essential ingredient of the *meaning* of "now"'.[2] The repeated use of the word 'meaning' seems to suggest that the mind-dependency of the present is based on an analysis of the meaning of terms rather than on any empirical considerations.[3] At any rate, if there are any empirical considerations which support this bald claim they are not presented.

At other places, however, what looks like an empirical consideration is advanced in support of the mind-dependency thesis. Both Bergmann and Grünbaum claim that physics takes no cognizance of the distinction between past, present and future, although it does take note of temporal relations between events. Grünbaum writes:

> It seems to me of decisive significance that no cognizance is taken of nowness (in the sense associated with becoming) in any of the extant theories of physics. If nowness were a fundamental property of physical events themselves, then it would be very strange indeed that it could go unrecognized in all extant physical theories without detriment to their explanatory success.[4]

This *appears* to be an empirical claim, for supposedly it is only a contingent fact about physics that it does not recognize the distinction between past, present and future, that its laws and theories make no reference to these *A*-determinations. This claim

[1] Grünbaum, *op. cit.*, p. 325. My italics.

[2] *Ibid.* My italics.

[3] Grünbaum's more recent article, 'The Temporal Status of Becoming', in *The Philosophy of Time*, edited by Gale, is more careful in its attempt to avoid the presentation of *a priori* arguments for the subjectivity of becoming, but it suffers a serious lapse when Grünbaum tries to support his thesis by an analysis of the meaning of a statement. He wrote: 'That the nowness attribute of an occurrence, when ascribed non-arbitrarily to an event, is inherently mind-dependent seems to me to emerge from a consideration of the kind of information which the judgment "It is 3 P.M., E.S.T. now" can be warrantedly held to convey . . .', pp. 335–336.

[4] Grünbaum, *ibid.*, p. 336.

by itself does not support the mind-dependency of *A*-determinations, but when it is combined with the unstated assumption that any objective, non-mind-dependent property should be taken note of by physics it does. This argument can appropriately be called the Argument from Scientific Irrelevancy; it holds for Newtonian science as well as for relativity theory.

What Bergmann and Grünbaum, I believe, have in mind when they claim that *A*-determinations are irrelevant to physics is that physicists have not been able to discover any physical basis or physical criterion for distinguishing some particular event or time as uniquely present—as *the* present. They agree with Hans Reichenbach who claimed that 'if there is Becoming the physicist must know it'.[1] Eddington, Reichenbach, Bondi and Whitrow have claimed to have found in the indeterministic laws of quantum physics a physical basis for becoming. *The* present is the divide between the recordable past and the indetermined future. This dispute makes it look like it is a contingent question whether there is, or could be, any physical criterion for picking out some event or time as the present. It will now be argued that the claim that physics cannot find any physical basis for singling out some event as the present is analytic, and therefore the argument from scientific irrelevancy is really *a priori*. Once this is accomplished the soundness of this *a priori* argument can be assessed.

Let us first consider the attempts to find a physical basis for becoming. Reichenbach has contended that the present is the moment at which what was undetermined becomes determined.

> . . . in quantum physics . . . there are future facts which cannot possibly be predicted, whereas there are no past facts which it would be impossible to know. In principle, they can always be recorded . . .
>
> The distinction between the undetermined future and the determined past has, in the final analysis, found expression in the laws of physics: the present, which separates the future from the past, is the moment at which that which was undetermined becomes determined, and 'becoming' has the same meaning as 'becoming determined'.[2]

[1] H. Reichenbach, *The Direction of Time* (Berkeley, 1956), p. 16.

[2] Reichenbach, 'Les Fondements Logiques de la Mecanique des Quanta', *Annales de l'Institut Henri Poincare*, Vol. XIII (1953), pp. 154–157. Quoted from p. 320 of Grünbaum's *Philosophical Problems of Space and Time*.

The criticism which Bergmann and Grünbaum make of this physical criterion for determining what is *the* present is that it fails to single out any event as uniquely present—as *the* present—for it holds true for every event in history.[1] At the time of Plato's birth, past and future events were those which were then recordable and unpredictable, respectively; but this holds for every event in history. Thus Reichenbach's criterion does not enable us to pick out one of these events as *the* present one: he has not given us a criterion for being *the* present but only for being *a* present. At best, his criterion enables us to distinguish between an earlier and later direction of time, which is no more than what the logical asymmetries between past and future of Chapters VII and VIII accomplished.

The Bergmann–Grünbaum criticism of Reichenbach's criterion is correct, and, what is more, would apply equally well to *any* physical criterion for distinguishing some event or time as *the* present. The reason is that a physical criterion will be based upon some physical law or theory. Any such law or theory is general: the statement of it is temporally unrestricted since it quantifies over all times. Since the law or theory holds true for *any* time it obviously cannot serve as a criterion for picking out some one time as *the* present: if it holds for all times it cannot hold for just one time. Therefore, it turns out that it is not a contingent fact that there is no physical criterion for determining what is *the* present. And, as a result, the argument from scientific irrelevancy for the mind-dependency of becoming is *a priori*, since it is based on the analytic truth that there cannot be a physical basis or criterion for *the* present.

Is this *a priori* argument sound? The argument, once again, is:

(1) *A*-determinations are not taken cognizance of by physics.
(2) Any non-mind-dependent property is taken cognizance of by physics. Therefore,
(3) *A*-determinations are not non-mind-dependent properties.

This syllogism is valid, but are its two purportedly analytic premisses true? The first premiss was interpreted to mean that there is no physical criterion or basis for determining which event is *the* present one, and it was argued that so interpreted this premiss is an analytic truth. However, there are other plausible

[1] See Grünbaum, *ibid.*, pp. 323–324.

interpretations of this premiss in which it is either false or highly questionable. Among these interpretations are:

(1*a*) Physicists make no use whatsoever of *A*-expressions in their scientific activities.

(1*b*) Physical laws and theories contain no concepts which involve *A*-determinations as part of their meaning.

If the first premiss is interpreted as (1*a*) it is obviously false, for it is an undisputed fact that physicists make ample use of *A*-expressions in their laboratory work in which laws are discovered and tested. Grünbaum would reply that we must distinguish between the pragmatic and semantic levels of physics.[1] The pragmatic level has to do with the discovery and verification of laws and theories, the semantic level with the statement of these laws and theories themselves. Grünbaum grants that *A*-expressions enter in on the pragmatic level of physics, but denies that they enter in any way into the formulation of laws and theories: they are irrelevant on the semantic level.

The claim that *A*-expressions do not enter into the semantic aspect of physics leads us into interpretation (1*b*) of the first premiss, according to which physical laws and theories contain no concepts which involve as part of their meaning the notions of past, present and future. As we saw, physical laws and theories make no explicit reference to *the* present (past and future) since they are temporally unrestricted; and it was for this reason that they could not serve as a criterion for *the* present. It would be granted by everyone that the concept of *earlier than* is involved in the semantic formulations of certain physical laws and theories. On page 92 it was shown that *earlier than* could be analysed in terms of a disjunction of five *A*-determinations as follows:

(II) P *is* earlier than *Q* .≡. *P* is past and *Q* is present *or P* is past and *Q* is future *or P* is present and *Q* is future *or P* is more past that *Q or Q* is more future than *P*.

If this analysis is correct it shows that *A*-determinations are presupposed on the semantic level of physics, and not just on the pragmatic level of discovery and verification.

Grünbaum might reply that this analysis of *earlier than* in terms

[1] See Grünbaum's 'Carnap's Views on the Foundations of Geometry', in *The Philosophy of Rudolf Carnap*, edited by P. A. Schilpp (LaSalle, 1963), p. 629.

of *A*-determinations works, if it works at all, only if we are talking about the ordinary or common-sense concept of *earlier than*; however, the concept of *earlier than* which figures in the semantic formulation of physical laws and theories is not the ordinary one.[1] Unquestionably the physicist's concept of *earlier than* is not the ordinary one, especially in regard to whether this relation determines a discrete or continuous order and whether it is a triadic relation. The question, though, is whether it is similar to, or shares features in common with, the ordinary concept of *earlier than*? If the physicist's concept of *earlier than* is totally dissimilar to the ordinary one it is misleading for him to use this term: better that he speaks of events as bearing 'temporal' relations of 'shmearlier than' to each other. My belief is that there is some common core of meaning between the ordinary and physical concepts of *earlier than*. If this is so we should expect the above equivalence to go through for the physical concept of *earlier than*, especially since it deals only with topological properties of this relation. If this equivalence does not hold for the physical concept then I am at a loss to know what the physicist means by his use of 'earlier than'. I realize that questions concerning the relation between the ordinary and physical concepts of time are highly mooted, and therefore I only want to draw the weak conclusion that if premiss (1) is interpreted in accordance with (1*b*) its truth is questionable.

Having shown that there are plausible renderings of premiss (1) on which it is either false or questionable, we can next consider premiss (2), which says that any non-mind-dependent property is taken cognizance of by physics. Obviously, premiss (2), since it contains the expression 'is taken cognizance of by physics', is subject to exactly the same ambiguity as is premiss (1). If the syllogism is not to contain four terms it is necessary to interpret this expression in the same way in premisses (1) and (2). If interpreted as it is in (1*a*) and (1*b*) the soundness of the syllogism is called in doubt, for premiss (1) is either false or highly doubtful, respectively.

Even if premiss (1) is interpreted to mean that there is no physical criterion for *the* present, in which case we would have to give a similar rendering of 'is taken cognizance of by physics' in

[1] For a discussion of this see Grünbaum's *Philosophical Problems of Space and Time*, Chapters 7 and 8.

premiss (2), the soundness of the argument is doubtful, the reason being that premiss (2) is itself doubtful. Why must any objective, non-mind-dependent property be taken cognizance of by physics?[1] That it must be certainly is not advanced as an empirical claim, but if it is an analytic claim on what conceptual truth does it rest? I can think of none. Premiss (2) could be challenged by the following counter-example: *A*-determinations are objective properties, and yet they are not taken cognizance of by physics (in the above sense). To rule out this counter-example without any further justification is to beg the very point at issue.

In summary: the *a priori* argument from scientific irrelevancy is valid but the truth of its premisses is doubtful. On certain plausible interpretations its first premiss is either false or doubtful; and even when interpreted in such a way that it is an analytic truth its second premiss is highly suspect. If the thesis that becoming is mind-dependent is to rest solely on the argument from scientific irrelevancy it has a shaky foundation indeed.

It might be claimed at this point that there is a sense which has not yet been explored in which modern science shows the mind-dependency of becoming, namely by showing that the conditions which must obtain for events to undergo becoming involve reference to a sentient organism. Thus, while science does not 'take cognizance' of *A*-determinations it does 'make allowances for' *A*-determinations, in the sense of explaining the conditions under which events perceptually appear to have these determinations. Grünbaum has suggested that an analogy can be drawn between the way in which science since the time of Galileo has supported the thesis that secondary qualities are mind-dependent and the way in which science (supposedly) can also show the mind-dependency of becoming.

> It is apparent that the becoming of physical events in our temporal awareness does not itself guarantee that becoming has a mind-independent physical status. Common sense color attributes, for example, surely *appear* to be properties of physical objects independently of our awareness of them and are held to be such by

[1] J. J. C. Smart plainly is assuming this when he wrote: 'If past, present, and future were real properties of events, then it would require explanation that an event which becomes present in 1965 becomes present at that date and not at some other (and this would have to be an explanation over and above the explanation of why an event of this sort *occurred* in 1965)' (*Philosophy and Scientific Realism*, *op. cit.*, p. 135).

> common sense. And yet scientific theory tells us that they are mind-dependent qualities like sweet and sour are. Of course, if physical theory claims that, contrary to common sense, becoming is not a feature of the temporal order of physical events with respect to earlier and later, then a more comprehensive scientific and philosophical theory must take suitable cognizance of becoming as a conspicuous characteristic of our temporal awareness of both physical and mental events.[1]

The fact that terms for secondary qualities check out as objective by the standards of ordinary language, since they are part of inter-subjective discourse, does not show that secondary qualities have a reality independent of their being perceived. For according to science, if I may oversimplify, a physical object is really a collection of unobservable particles which do not possess any secondary qualities. Light waves are emitted from them, some of which impinge upon the retina of our eye. This causes an electro-chemical impulse to travel along the optic nerve to the occipital lobe in the brain, at which time we are visually aware of a coloured patch. It is not my purpose here to investigate the adequacy of this familiar causal theory of sense perception, but rather to inquire whether there is any analogous causal theory of the perception of becoming which is licensed by science.

As far as I can make out, no defender of the mind-dependency thesis has advanced anything which is even remotely analogous to the above causal theory of perception. That they have not done so is not a result of laziness or an oversight on their part, but rather is due to a basic difference between *A*-determinations and secondary qualities which makes it impossible to advance an analogous causal theory of the perception of becoming. The crucial difference is that *A*-determinations are not sensible properties, whereas secondary qualities are. Chapter V showed that *A*-expressions do not have the logic of sensible terms. We do not perceive *A*-determinations through one or more of our senses. For this reason a causal theory of the *perception* of *A*-determinations cannot get off the ground, since it must begin with *perceived*, i.e. sensible, properties.

It might be claimed that the causal theory of perception, by showing the mind-dependency of secondary qualities, thereby also establishes the mind-dependency of becoming. What argu-

[1] Grünbaum, 'The Status of Temporal Becoming', *op. cit.*, pp. 322–323.

ment could be presented for this? Certainly the contingent fact that our perceptual awareness of an event is later than this event is irrelevant. Equally irrelevant is the fact, if it is a fact, that physical objects and events do not have the sensible properties they appear to have in sense experience. It might be urged that since the causal theory of perception based on modern science has discredited our common-sense realism concerning sensible properties it also invalidates our common-sense commitment to the objectivity of becoming. Even if it were true, though it is by no means clear that it is, that science has discredited the realistic commitments of our ordinary discourse about sensible properties it does not follow that it has thereby also discredited the realistic commitments of our ordinary discourse about *A*-determinations. This would follow only if it could be shown that there is some logical connection between the logic of sensible terms and *A*-expressions such that if the realistic commitments of one of these is discredited so too is that of the other. It is very unlikely, in view of the fact that *A*-expressions are not sensible terms, that this could be done. The scientifically grounded causal theory of perception not only does not establish the thesis that becoming is mind-dependent but, as we shall soon see, it raises serious difficulties for this thesis.

Having shown so far that there are no *a priori* or empirical arguments which can establish the mind-dependency of becoming, we next turn to a consideration of the adequacy of this manner of representing becoming. It will be recalled that the mind-dependent thesis of Weyl, Eddington and Grünbaum represented becoming as due to the mind crawling or voyaging up a world-line in which it comes upon pre-existent physical events one after the other.

> The objective world simply *is*, it does not happen. Only to the gaze of my consciousness crawling upward along the life (world-) line of my body does a section of this world come to life as a fleeting image.[1]

> Events do not happen; they are just there, and we come across them. 'The formality of taking place' is merely the indication that the observer has on his voyage of exploration passed into the absolute future of the event in question; and it has no important significance.[2]

[1] Weyl, *op. cit.*

[2] Eddington, *op. cit.*

These representations are variations on the River of Time picture of becoming, according to which physical events are like flotsam which are swept downstream by the River of Time and acquire presentness as they pass through a spotlight which an observer on a bridge is shining down on to the water. These physical events are substantialized because they *co-exist* in a one-dimensional *spatial* order and *endure* in the same spatial order. The above are representations of the Frozen River of Time. Physical events are like snow men lined up in an unchanging spatial order on the frozen River of Time: the observer—the mind—is no longer standing idly on a bridge watching the passing parade of events but is voyaging, possibly on iceskates, up this river and comes upon these 'pre-existent' snow-man-events one after the other. These snow-man-events, unlike real snow men, can never melt, for events cannot change either in any of their intrinsic properties or in their temporal relations to other events. If they could melt there would be future contingents—snow men that might melt before our mind reaches them: instead of coming upon the death of Queen Anne we might just step or skate into a puddle of water. Of course, the issue of determinism is completely irrelevant to the question concerning the objectivity of becoming.

It is obvious that the frozen version of the River of Time involves the same conceptual absurdities as the flowing River of Time. In both versions there is a substantialization of events and consequent spatialization of time. Since earlier and later events are co-existent substances which endure there must be a meta-time—a fifth dimension—for this endurance, this remaining unchanged, to occur in. Also, the concepts of flowing and voyaging involve a change of spatial position over an interval of time; but in these representations what the river and the mind respectively flow and voyage along is an interval of time. This is absurd, and its absurdity is independent of whether or not time and space have been metricized or even are metricizable.

Today these Frozen River of Time representations of becoming are generally recognized as metaphors whose literal meaning is as follows. Physical events are not in themselves past, present or future: they only *are* earlier and later than and simultaneous with other events. This is the literal meaning of 'the objective world simply *is*', and 'Events . . . are just there'. A physical event acquires the property of presentness—becomes present, happens, takes place—only when a mind 'comes across' it, that is when either it or some physical event simultaneous with it is perceived by a mind.[1] A physical event is past (future) only in the elliptical sense of being earlier (later) than some mental event.[2] Thus there is a series of mental events which runs parallel to the series of physical events, events in the latter series aquiring the distinction of being future, present and past only in virtue of being respectively later than, simultaneous with and earlier than some event in the former series.

The becoming of physical events is explained in terms of a mind's successive perception of these events, as seen in this diagram:

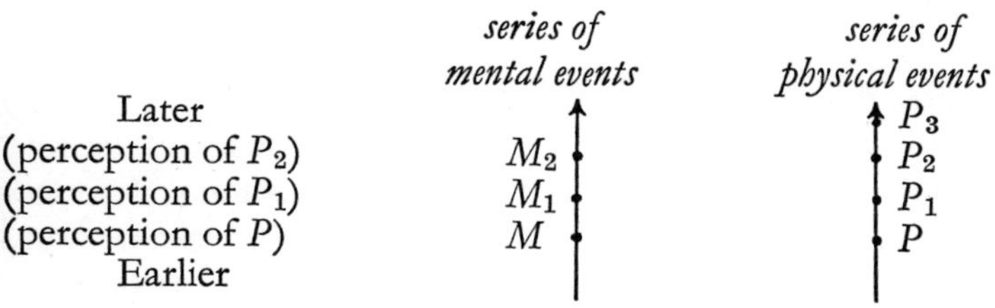

P, P_1 and P_2 undergo becoming because they are successively perceived respectively by M, M_1 and M_2. The becoming of physical event P_1—its being future, then becoming present and then past—is analysed in terms of the fact that it is later than mental event M, simultaneous with a subsequent perception of it, namely M_1, and earlier than a later mental event M_2. Since the events in these two series of events bear *B*-relations to each other, they

[1] Grünbaum adds the further requirement that this perception be one of which the percipient is self-consciously aware. In 'The Status of Temporal Becoming', p. 333, he said: 'What qualifies a physical event at a time *t* as belonging to the present or as now is not some physical attribute of the event or some relation it sustains to other purely physical events; instead what so qualifies the event is that at least one human or other *mind-possessing* organism *M* experiences the event at the time *t* such that at *t*, *M* is *conceptually aware* of the following complex fact: that his having of the experience of the event *coincides temporally* with his awareness of the fact that he has it at all.' Grünbaum gives no justification for adding this further requirement of self-conscious awareness to the earlier psychological theories of presentness. We shall not consider his more refined analysis, but all of the objections which will be raised against the psychological representation of becoming apply equally to his account.

[2] In some accounts, such as the one on page 21, a physical event is future only if later than an expectation, and is past only if earlier than a memory experience, in which the expectation and memory need not be of this physical event. We shall overlook this further requirement for physical events being future and past. It will make no difference in our discussion.

combine to form a single *B*-Series. Notice, incidentally, that physical event P_3 does not acquire the property of presentness since it is not perceived or simultaneous with a perception, although it does acquire the property of futurity when mental event M occurs, as well as M_1 and M_2.

The psychological representation of becoming, even when purged of misleading metaphors as it has been above, still involves serious conceptual difficulties. One type of difficulty concerns its theory of mind, and the second its conception of the tenseless being of physical events. Concerning the first type of difficulty it will be argued that this representation is committed to a strange theory of consciousness which makes the mind into a ghost having no connection, be it causal or logical, with the body.

The first thing to note about the psychological representation of becoming is that it involves a rigid dichotomy between mental and physical events. This dichotomy, although it is not without its own difficulties, will be accepted for the sake of argument. It will now be shown that the psychological representation is committed to a bizarre 'non-naturalistic' theory of mind because: (i) it is unable to account for the interaction between mind and body; and, (ii) leaves us with no criterion, either causal or logical, for identifying the occurrence of mental states.

(i) According to the psychological representation there is a series of mental events which runs parallel to the series of physical events, such that events in the latter series undergo temporal becoming only in virtue of their relation to the events in the mental series. A physical event becomes present only in the elliptical sense of being perceived; the becoming of physical events is due to the successive perceptions of them by some mind. These mental events bear unchanging *B*-relations to each other, and thus enjoy the same sort of tenseless existence as do physical events. The vital question concerns whether these mental events, in addition to having a tenseless existence in the *B*-series, also *intrinsically* become? Are mental events, such as perceptions, present *simpliciter*? Or are they, like physical events, present only in a relational way?

It is an uncontestable fact that our mental states *appear* to undergo becoming: right now one or more of the mental states which make up our mental history is occurring, while the others are past or future; and moreover this present mental state will become past as some new mental state becomes present. This temporal becoming of our mental states cannot be accounted for, as Grünbaum thinks it can be, solely in terms of 'a *diversity* of Now-contents of immediate awareness';[1] for physical events occurring at different times also differ in their contents, and yet they are not said to become intrinsically, that is to be present in some non-relational sense and change in respect to their *A*-determinations. There is not only a diversity in the contents of our acts of awareness from time to time but also these acts of awareness themselves appear to undergo becoming.

If these appearances are to be saved, it must be granted that mental events are intrinsically past, present and future and change in respect to these *A*-determinations. It cannot be claimed, without vicious circularity, that a mental event, such as an act of perceptual awareness, is present only in the elliptical sense of being related to some physical event of which it is a perception; for it has already been contended that a physical event is present only in the elliptical sense of being the object of some perception. Nor can it be claimed, without instituting a vicious infinite regress, that a mental event is present only in the elliptical sense of being related to some mental event in a second-order series of mental events. For the same question can be raised in respect to this second mental event, 'Is it present simpliciter?', and so on *ad infinitum*. This infinite regress is vicious because it makes it impossible to account for the uncontestable fact that events appear to be present. Therefore, the events in the series of mental events must be past, present and future simpliciter.

Similar considerations would apply to a linguistic representation of becoming, according to which physical events become present only in the elliptical sense of being simultaneous with an *A*-sentence token which reports their occurrence. The events in the series of linguistic events must be present simpliciter, for the reasons given above.[2] A linguistic representation of becoming lacks the plausibility of a psychological one, since linguistic

[1] Grünbaum, *Philosophical Problems of Space and Time*, p. 315.

[2] Even on the TRI analysis of *A*-statements there still is an entailment about an event—a physical one—being present simpliciter. The statement in the *analysans* that *X is* simultaneous with this token entails that this token is now occurring. This is a result of the fact that 'this' means what is being pointed at or referred to at present.

events are themselves physical events. If linguistic events can be present simpliciter why cannot other physical events, even though non-linguistic? At least on the psychological representation the events which intrinsically become are mental events, which are conceded to have some very peculiar metaphysical properties, such as not existing in space.

Granted, as it must be, that mental events have *A*-determinations in a non-relational manner, how can the psychological representation account for their causal connection with physical events, which, according to this representation, enjoy only a tenseless existence? A naturalistic theory of mind must be able to give such an account. Let us consider the problem of physico-psychical causation. There is overwhelming scientific evidence that states of perceptual awareness, along with other mental states, are triggered or caused by brain events: the causal theory of perception is based in part on such scientific findings. A brain event is, of course, a physical event, and thus cannot be present simpliciter on the psychological representation; however, a simultaneous mental event which is causally dependent upon this physical event does become present simpliciter. This is most strange. Since mental and physical events are temporally, although not spatially, related to each other, we should expect mental events to enjoy the same sort of *temporal* existence as do the simultaneous physical events which trigger them. On the psychological representation the mind is a ghost which in some mysterious and unexplainable fashion successively 'comes upon'—perceives—different time-slices of the four-dimensional physical world, generously bestowing upon each of these time-slices in turn the gift of presentness. Thus, the causal theory of perception, rather than giving any scientific support to the psychological representation of becoming, creates serious problems for this representation.[1]

The inability of the psychological representation to account for physico-psychical causality also renders mysterious the temporal relations between mental events. Temporal relations between mental events are determined by the order in which these events

[1] Psychico-physical causation would also be rendered mysterious on the psychological representation. A mental event *M* which is the cause of a later physical event *P* would happen but it would not make *P happen*, simply because *P*, being a physical event, does not happen. This point is suggested by H. H. Price's 'The Philosophical Implications of Foreknowledge', *Aristotelian Society Supp.*, Vol. 16 (1937), p. 227.

become present. Since physical events do not become present simpliciter there can be no physical explanation for why mental events become present in the order they do, and therefore no explanation for why these events bear the temporal relations to each other which they do. There is no reason why the temporal relations between mental events should not be the reverse of the temporal relations between the physical events which 'cause' them, or why our minds should not be able to shift about in time and perceive events in any arbitrary order. Nor is there any reason why a physical event should not be perceived more than once by a single mind, and as a result become present on more than one occasion.

C. J. Ducasse, one of the proponents of the psychological representation, accepts this spooky consequence, and, in fact, utilizes it to account for how precognition is possible without backward causation.

> What really occurs there (precognition) is only that one individual physical event, although occurring of course only once, nevertheless is present several times because perceived at several psychological times—once perhaps, at night in a dream; and again, perhaps the next day, in ordinary waking perception. That it can be present several times without itself occurring several times is an automatic consequence of the cardinal fact that, in itself, it is neither present nor past nor future but acquires presentness by being perceived and as often as it is perceived.[1]

The mind, so to speak, can go A.W.O.L. from the moment of time to which it is supposed to be stationed, can sneak up the time-axis, perceive a future event and then slip back to camp without ever having been missed. This way Ducasse can explain precognition without violating the analytic truth that a cause cannot be later than its effect.

Most of the defenders of the psychological representation, such as Grünbaum, would not welcome this spooky consequence with the enthusiasm of Ducasse; for they have tended to be naturalists in their theory of mind, believing that consciousness is causally dependent upon the body.

(ii) Another bizarre result of the doctrine that mental, but not

[1] C. J. Ducasse, 'Broad on the Relevance of Psychical Research to Philosophy', in *The Philosophy of C. D. Broad*, *op. cit.*, p. 393.

physical, events are present simpliciter is that it leaves us with no criterion, causal or logical, for identifying the occurrence of different kind of mental events and states. Our criterion for determining the occurrence of a certain kind of mental event is based upon physical events, i.e. publicly observable events. By definition we cannot experience the mental states of other persons, so our only criterion for identifying the occurrence of their mental states is based on their bodily behaviour, which includes their linguistic avowals. This is true regardless of whether we view this criterion as causal or logical. It is causal if we take the connection between mental and physical events to be contingent; it is logical, however, if we take this connection to be a necessary one, necessary in that it is based on the meaning or logic of psychological words. Someone who views the criterion in the latter manner need not deny that mental events are numerically distinct from physical ones.

If only mental events become present simpliciter we are robbed of this criterion for identifying what kind of mental events are happening. If, in other words, we can identify physical events as present only by first identifying some mental event as present, we no longer have any criterion for identifying the present occurrence of different kind of mental events. The physical events which we observe, such as bodily behaviour, are not present simpliciter; and as a result we can have no assurance that the mental states with which they would ordinarily be said to be causally or logically connected become present at the time of our observations. For example, I might now observe a man in front of me doubled over, bleeding and screaming that he is in pain; since these physical events are not present simpliciter, but only in relation to a mental event, I have no basis for claiming that this person is now in pain, that is that his *feeling* of pain is now present simpliciter.

Because of the lack of any criterion for identifying the occurrence of mental events, we have no criterion for determining when the mental states of different minds are simultaneous. This has serious social implications, as the following example brings out. Jones and Smith meet at the smoker of the American Philosophical Association Meetings and shake hands. At this time there occur two almost coincident physical events, P_j and P_s, consisting in the contraction of the muscles in the right hands of Jones and

Smith respectively. P_j and P_s are simultaneously accompanied by mental events M_j and M_s which consist respectively in the conscious awareness of P_j and P_s. Are M_j and M_s simultaneous? They are tenselessly simultaneous respectively with P_j and P_s, which themselves are tenselessly simultaneous with each other; and thus M_j and M_s are tenselessly simultaneous. But this does not answer our question. Because mental events become present simpliciter, two mental events are simultaneous only if they become present simultaneously. As argued above, P_j cannot serve as a criterion for determining that M_j becomes present, and similarly for the relation between P_s and M_s. There is no criterion, therefore, for determining that M_j and M_s simultaneously become present, and thus no criterion for their simultaneity. It is possible, for all we could know, that when M_s becomes present M_j has already done so. In other words, by the time Smith's mind 'comes upon' this handshake the mind of Jones may have beaten him to it and already be perceiving physical events in the absolute future of this handshake event. This may account for the mysterious, far-off look of people you shake hands with at convention smokers.

The psychological representation, in addition to being committed to a strange non-naturalistic theory of consciousness, also involves the absurdity of there being physical events which exist in only a tenseless manner. Weyl and Grünbaum both say that physical events 'simply are' or 'simply have being', Eddington that they 'do not happen' or 'take place'. This is contradictory, for by definition an event happens, takes place or occurs. Moreover, what happens becomes present. Therefore, for any event it is true that either it was present or is now present or will be present. The physical 'events' of the psychological representation do not satisfy this analytic requirement for being an event.

It is claimed by the psychological representation that the 'events' in the series of physical events are tenselessly related to each other as well as to those in the series of mental events. It is an analytic truth that temporal relations of precedence, subsequence and simultaneity can hold only between *events*, i.e. things that happen. Since physical 'events' do not happen they are not events, and therefore cannot bear any temporal relations either to each other or to mental events. This means that Smith's conscious awareness of the handshake cannot be simultaneous with this physical event.

Another difficulty with speaking of there being temporal relations between events which do not have *A*-determinations is that our ordinary concept of a temporal relation involves *A*-determinations. Chapter VI showed that a definition in use could be given of *B*-relations in terms of the five *A*-determinations. Therefore, it is contradictory to speak of events as bearing *B*-relations to each other but not having *A*-determinations. Once again, it might be urged that the *B*-relations between physical events are not the ordinary ones; but this only raises the questions 'Why aren't these relations the ordinary ones?' and 'If they are not then what are they?—*shmearlier than* and *shlimultaneous with* maybe?'

Probably the most patent conceptual absurdity in the view that physical events become present only when perceived (or simultaneous with a perceived event) is the following. According to the psychological representation a physical event is future only if later than some mental event. It is possible that a future physical event, such as P3 in the diagram on page 231, will not be perceived (nor be simultaneous with a perceived event), and therefore will not become present. But a future event is by definition an event which will become present. Therefore, it is possible that there is an event which will become present and which will not become present.

That the psychological representation is absurd when judged from the standpoint of common sense or ordinary language does not thereby discredit it or establish its falsity. By now it is a familiar cautionary tale that many of the great conceptual innovations of modern science were judged to be absurd by the canons of the then prevalent system of common-sense concepts. This, however, is of no avail to the psychological representation; for, as argued at length in this chapter, this view finds no support in any empirical discovery of science nor in any feature of physical theory. Quite the opposite was found to be the case, the non-naturalistic theory of mind which this view is committed to being inconsistent with a vast array of empirical evidence that consciousness is causally dependent upon the body.

There is only one final move—a truly desperate one—open to one who believes that becoming is not an objective feature of reality. This is to claim that this concept is itself absurd and therefore cannot be true of reality. An Eleatic might say that what is the case cannot cease to be the case; and therefore the concept of becoming is contradictory. But the Eleatic principle appealed to is nothing but the mystic's claim that Being is timeless since unchanging. There is no evidence or arguments to support this—quite to the contrary. Another example of a desperation move to show the impossibility of there being becoming is McTaggart's argument. It is to be recalled from Chapter I that McTaggart claimed that the *A*-Series is contradictory since every event in it has all three mutually incompatible *A*-determinations of past, present and future. And when we attempt to explain away this seeming contradiction by saying of some event, for example *M*, that it was future, is now present, and will be future, we become involved in either a vicious circle or a vicious infinite regress. The reason is that we must *analyse* this statement into '*M is* at future a moment of time which is past, *is* present at a moment of time which is present, and *is* past at a moment of time which is future', and thereby invoke moments of time in some second-order *A*-Series: and the same contradiction breaks out afresh, since each of these moments of time has all three *A*-determinations.[1] The way to break the back of this argument is to deny that this analysis of *A*-statements is correct.

Stebbing and Wisdom, as seen on page 31, used this way out by claiming that 'M is present' entails, but does not mean, 'M *is* present at a moment of time which is present'; and as a result McTaggart's infinite regress of meta-times-series is benign since it is a regress of entailings and not meanings. Two things trouble me about this way out. First, a prominent theory of entailment holds that an entailment depends on a connection of meanings; and if this theory is correct the regress is one of meanings and therefore vicious. I do not say that this is the correct view; but it would be preferable to refute McTaggart without begging any mooted questions. Second, why should that M is present entail that there are moments of time. It certainly seems conceivable for M to be present without there being such strange entities as moments of time.

[1] McTaggart, in *The Nature of Existence*, *op. cit.*, p. 21, wrote: 'When we say that *X* has been *Y*, we are asserting *X* to be *Y* at a moment of past time. When we say that *X* will be *Y*, we are asserting *X* to be *Y* at a moment of future time. When we say that *X* is *Y* (in the temporal sense of "is") we are asserting *X* to be *Y* at a moment of present time.'

That McTaggart's analysis of *A*-statements is incorrect can be seen by reference to the following paradigm case of tensed communication. *A* and *B* are standing face-to-face and *A* says to *B*, 'She is on her way over now'. To *A*'s surprise *B* then asks, 'But is it now now?' or 'When is she on her way over now?' The appropriate answer to *B* would be to inquire whether he understands English for he has just been told when she is on her way over, i.e. right now. Certainly we would not answer him by saying that she is on her way over at a moment of time which is now present. Because B's question is pointlessly redundant, McTaggart's analysis of 'M is present' into 'M *is* present at a moment of time which is present' is incorrect; for according to his analysis *B*'s question would not be redundant, and the answer to it would be that she *is* on her way over at a moment of time which is present. But in accordance with this analysis, we are justified in coming back with an identical question concerning the moment of time—when is it present?—and so as to infinity. However, this hopeless infinite regress of meta-time-series can be nipped in the bud when we see the pointless redundancy of *B*'s initial question. McTaggart's analysis of 'M is present' into 'M *is* present at a moment of time which is present' suggests that there is a second-order time-series composed of moments of time at which events in the first-order time-series are present. This is to dive into the muddy waters of the River of Time, the essence of which is to think of *the* present moment as some point or observer 'outside' of the first-order time-series. McTaggart's transcendent entity *X* in his subsidiary argument (on page 12) is a transcendent spotlight of presentness. Events in this first-order time-series receive their *A*-determinations in virtue of their relation to this transcendent point, which 'moves' along this time-series.[1]

The error which underlies McTaggart's analysis of *A*-statements is the obsessive belief, common to defenders of the *B*-Theory, that *A*-determinations are *relations*, or, in other words, that *A*-expressions are disguised two-place predicates. In the last chapter it was shown that that *M* is present does not entail that

[1] While it is meaningless to ask when a present event is present it is not meaningless to ask when a past (future) event was (will be) present, the answer to which might employ metrical concepts, such as 'seven days ago'. But certainly our use of these metrical concepts does not commit us to accepting moments of time.

there is any mental or linguistic event simultaneous with *M*, and for this reason the relational analyses of EP and TRI fail. It certainly does not entail that there is a moment of time simultaneous with *M*. Moreover, that *M* is present does not even entail that there is any event simultaneous with *M*, and for this reason this statement cannot be analysed into '*M is* simultaneous with this', in which 'this' denotes any event other than *M* itself.

Another way of bringing out the fact that *A*-expressions are not disguised two-place predicates is as follows. A statement containing a disguised two-place predicate, such as '*X* is a father', can be expanded into an explicitly relational statement, such as '*X* is the father of ____', in which some singular or general expression referring to a person is substituted for the blank space. The expression substituted for the blank space need not contain as part of its meaning the notion of *being a father*, which is the disguised relational term. For instance, if 'some person' is substituted for the blank space this expression does not contain as part of its meaning the notion of being a father. If we treat 'is now present' as a disguised two-place predicate we would have to analyse '*X* is now present' into '*X is* simultaneous with ____', in which some expression referring to an event is substituted for the blank space. If the parallel is to hold, the expression which is substituted for the the blank space need not contain as part of its meaning the notion of being present, which is the disguised relational term. But in order for the latter, explicitly relational statement to convey at least as much information as conveyed by the former, we would have to substitute for the blank space an expression such as 'this' or 'now': '*X is* simultaneous with this (now).' However, the meaning of the expression substituted for the blank space contains the notion of being present, which is the supposedly disguised relational term. *Webster's New Collegiate Dictionary* defines 'this' as: 'A demonstrative word, referring particularly to what is *present*. . . .'

There is another move open to one who believes that becoming is an absurd concept. It might be argued that if events become—change in respect to their *A*-determinations—they must do so at a certain rate, so many events or units of time becoming present in a unit time. This presupposes a meta-time series and thereby institutes a vicious infinite regress of times changing within times. Therefore, events do not undergo becoming. This argument assumes that becoming is a non-trivial metrical concept, that is, a question about the rate at which events temporally become cannot be answered by an analytic statement, as for instance 'Events change in their *A*-determinations at the rate of one hour per hour'. The assumption that becoming is a metrical concept in some non-trivial sense is faulty, and I should now like to indicate just why.

It is only in regard to a change *in* time—a change in the intrinsic or relational properties of a thing (substance)—that it is meaningful to ask in some non-trivial sense, 'At what rate does this change take place?' Such a question is answered by correlating the different stages or events in this change with the changes in another thing which serves as our clock. Since time, as well as its parts, such as past, present and future, is not a thing or continuant it is meaningless to say that time changes at a certain rate—that the present shifts at a certain rate. Becoming is a change *of* time and cannot without absurdity be reduced to a change *in* time. Becoming involves a change in regard to the *A*-determination of events. Since events are not things it cannot be asked, 'At what rate do events become?' To ask this question is to accept a River of Time representation of becoming, with its vicious substantialization of events and spatialization of time. Any attempt to reduce change *of* time to change *in* time must lead to absurdities. This is due to the fact that the concept of becoming is pre-supposed by the concept of a change *in* time. The concept of change *in* time involves *events* happening to a single thing at different times; but the concept of an event is in turn analysed in terms of becoming present—becoming. Since becoming is presupposed by change *in* time obviously we cannot say and ask the same things about becoming as we can about the latter.

The upshot of this discussion is that becoming is a *sui generis* type of change. Failure to realize what is unique about time has often led to paradoxical results, of which the above argument is a paradigm example. Paradox will always result from applying temporal concepts to time. Many supposed aphorisms about time are guilty of this. It is often said that time is *continually* or *always* passing, that the present *continually* shifts, or that new events are *always* becoming present. But it is only because new events become present that it is meaningful to say of some thing that it *continually* or *always* changes or remains the same. For this reason it is meaningless to speak of new events as always or continually becoming present or of the present as continually shifting. To speak in this way is to presuppose a meta-series of events which undergo becoming, which is the first step of the vicious infinite regress. What time makes it possible for us to say is exactly what cannot be said about time.

INDEX

Index

Pcl#: 14136543 Typ:R4 Tm: WPK545 Order # 60702234SM Ctn: 1 of 1

Pl#: 1098 Date: 8/30/03 Wve:001 Tote: 14154296 **SLA: 5**

Location	ISBN	Pieces	Description
545-122-021	0534464505	1	2004 WW PHIL/REL CATALOG

The enclosed materials are compliments of
BRAD SCHADLER

#00003

SHC: 0 THOMSON LEARNING

Page: 1 of 1

International Library of Philosophy & Scientific Method

Ted Honderich

o)

. (Ed.), **Studies in Plato's Metaphysics** *464 pp. 1965.*
s **'Euthyphro' and the Earlier Theory of Forms** *184 pp. 1970.*
. and Furley, David J. (Eds.), **Studies in Presocratic Philosophy**
1970.
, D.M., **Perception and the Physical World** *208 pp. 1961.*
erialist Theory of the Mind *376 pp. 1967.*
gh, Renford (Ed.), **New Essays on Plato and Aristotle**
1965.
n, **Political Argument** *382 pp. 1965.*
am, **Kant's Theory of Knowledge** *220 pp. 1962.*
mes, **Wittgenstein's Philosophy of L**
D., **Le**
pres
. M., **n Examination of Plato's** **octrine**
on M n and Society *408 pp. 1962.*
on Ki owledge and Reality *583 pp.* *63.*
Patri k, **Inductive Probability** *35 pp. 1961.*
. C., **ontent and Consciousnes** *202 pp. 1969.*
red I., **Seeing and Knowing** *270 p* *1969.*
C. J., **T uth, Knowledge and Caus tion** *263 pp. 1969.*
ham, **ethod in Ethical Theory** *3 pp. 1963.*
(Ed.), **Symposium on J. L. Austi** *12 pp. 1969.*
hony, **ume's Philosophy of Beli** *296 pp. 1961.*
obert J. **Evidence and Meaning** *2 pp. 1967.*
., **Free will and Determinism** *35*
ard, **Th**
nathan
Lucien, **The Hidden God** *424 pp. 19 4.*
. W., **Se sation and Perception** *22 pp. 1961.*
pressio 1967.)
dmund, **ogical Investigations** *Vol. 456 pp. Vol. II: 464 pp.*
Reason Action and Morality *216 p* *1964.*
ephan, **E perience and Theory** *272* *. 1966.*
z, Morris **Studies in Metaphilosoph** *276 pp. 1964.*
onard, **R ferring** *152 pp. 1967.*
, J. J. an Coval, S. C. (Eds.), **Busines of Reason** *280 pp. 1969.*
ack W., **alking About Particulars** *2 pp. 1970.*
Ponty, M. **Phenomenology of Perce ion** *487 pp. 1962.*
e, **Scept cism** *176 pp. 1969.*
Chaim, **T e Idea of Justice**
1963.
Directive
er, G., **Me**
. F., **Scien**
Science, Percep
Shwayder, D. S., **The**

STATES AVENUE
WESTIN VACATION
STAMP 607
NEED STAMPS 606, 607 & 608 TO WIN

G
Del Monte
Quality
#4011

Thank you,
Mr. Hughen,
for remembering
America's heroes!

Good Sam Club

One
the better banana co.
RAINFOREST ALLIANCE CERTIFIED
4011
Guatemala
7404003900047
4011 One Banana

Dial
10811
+ 1+ Area Code+ Number
DIME LINE
CUSTOMER SERVICE 1-800-544-1510
VarTec Telecom, Inc.

Nations Photo Lab
11103 PEPPER RD

HUNT VALLEY, MD 21031-1203

NATIONS PHOTO LAB

Order Id: WEB11932200X01

DESCRIPTION	QTY
Include White Envelopes for 5x7 Cards (Pack of 26)	1
5x7 GC Blank L (Pack of 25) [Semi-Gloss] [CC]	1

Carol Hughen
3281 W Gypsum Dr.
Fayetteville, AR 72704-5610

Excellence Guaranteed

If there's something not quite right with your order, please let us know and we'll work hard to fix it. Simply email us at info@nationsphotolab.com or call us at 1-888-507-5755 to speak with a friendly customer service representative today

SHIP TO :
Hughen, Rich 1469292
3281 W GYPSUM DR

FAYETTEVILLE, AR 72704-5610

WEB11932200X01